THE THEORY
OF
ALGEBRAIC NUMBERS

HARRY
POLLARD
AND
HAROLD G.
DIAMOND

THIRD, REVISED EDITION

DOVER PUBLICATIONS, INC.
Mineola, New York

Bibliographical Note

This Dover edition, first published in 1998, is a lightly revised but otherwise unabridged republication of the work originally published in 1975 by the Mathematical Association of America (Incorporated) as the ninth in the series of The Carus Mathematical Monographs.

Library of Congress Cataloging-in-Publication Data

Pollard, Harry, 1919–
    The theory of algebraic numbers / Harry Pollard & Harold G. Diamond. — 3rd, rev. ed.
        p.   cm.
    Includes bibliographical references and index.
    ISBN 0-486-40454-4 (pbk.)
    1. Algebraic number theory.   I. Diamond, Harold G., 1940–
II. Title.
QA247.P6   1998
512'.74—dc21   98-28912
                                                                                  CIP

Manufactured in the United States by Courier Corporation
40454405   2013
www.doverpublications.com

DEDICATED TO
PROFESSOR DAVID VERNON WIDDER

PREFACE TO FIRST EDITION

The purpose of this monograph is to make available in English the elementary parts of classical algebraic number theory. An earlier version in mimeographed form was used at Cornell University in the spring term of 1947–48, and the present version has accordingly profited from the criticisms of several readers. I am particularly indebted to Miss Leila R. Raines for her painstaking assistance in the revision and preparation of the manuscript for publication.

PREFACE TO SECOND EDITION

This new edition follows closely the plan and style of the 1st edition. The principal changes are the correction of misprints, the expansion or simplification of some arguments, and the omission of the final chapter on units in order to make way for the introduction of some two hundred problems. The credit for this revision is almost entirely due to my associate, Harold G. Diamond.

We are grateful to many contributors for both correc- tions and problems: P. T. Bateman, B. C. Berndt, P. T. Montague, Ivan Niven, J. Steinig, and S. V. Ullom.

HARRY POLLARD

August 1974

PREFACE TO THE DOVER EDITION

This is a reprint of the Second Carus Edition, which remained in print until 1997. The present edition contains a corrected proof of Theorem 8.13 and alterations of a few of the problems.

Professor Harry Pollard, the author of the first edition, died in 1985.

Harold G. Diamond
February 1998

CONTENTS

THE THEORY
OF
ALGEBRAIC NUMBERS

Chapter I

DIVISIBILITY

1. Uniqueness of factorization. Elementary number theory has for its object the study of the integers 0, ± 1, ± 2, ... Certain of these, the *prime* numbers, occupy a special position; they are the numbers m which are different from 0 and ± 1, and which possess no factors other than ± 1 and $\pm m$. For example 2, 3, -5 are prime, whereas 6 and 9 are not, since $6 = 2 \cdot 3$, $9 = 3^2$. The importance of the primes is due to the fact that, together with 0 and ± 1, all the other integers can be constructed from them. The fundamental theorem of arithmetic asserts that *every integer greater than 1 can be factored in one and only one way, apart from order, as the product of positive prime numbers.* Thus

$$12 = 2^2 \cdot 3 = 2 \cdot 3 \cdot 2 = 3 \cdot 2^2$$

are the only factorizations of 12 into positive prime factors, and these factorizations all yield precisely the same factors; the only difference among them is in the order of appearance of the factors.

We shall give a proof of the fundamental theorem of arithmetic. In the course of it the following fact will play a decisive role: every collection, finite or infinite, of non-negative integers contains a smallest one. The validity of this assumption will not be debated here; it is certainly clear intuitively, and the reader may take it to be one of the defining properties of integers. Some preliminary theorems will be established first.

1

THEOREM 1.1. *If a and b are integers, b > 0, then there exist integers q and r such that*

$$a = bq + r,$$

where $0 \leq r < b$. The integers q and r are unique.

Consider the rational number a/b and let q be the largest integer which does not exceed it. Then $q \leq a/b$, but $q + 1 > a/b$. Define r as $a - bq$. Since $r/b = (a/b) - q \geq 0$, and $b > 0$, it follows that $r \geq 0$. Also from $1 > (a/b) - q = (a - bq)/b = r/b$ we conclude that $r < b$.

To show that q and r are unique suppose that q' and r' is any pair of integers for which

$$a = bq' + r', \qquad\qquad 0 \leq r' < b.$$

If $q' > q$, then $q' \geq q + 1$, so that

$$r' = a - bq' \leq a - b(q + 1) = r - b < 0;$$

this contradicts $r' \geq 0$. If $q' < q$, then $q' \leq q - 1$, so that

$$r' = a - bq' \geq a - b(q - 1) = r + b \geq b;$$

this contradicts $r' < b$.

Then both possibilities $q' > q$, $q' < q$ are ruled out. It follows that $q' = q$, and hence that $r' = r$. This completes the proof of Theorem 1.1.

We shall say that two integers a and b are *relatively prime* if they share no factors except ± 1. Thus 5 and 9 are relatively prime, whereas 6 and 9 are not.

THEOREM 1.2. *If a and b are relatively prime then there exist integers s and t for which $as + bt = 1$.*

Observe that there is no assertion about the uniqueness of s and t. In fact if $a = 3$, $b = 5$ we have

$$2 \cdot 3 - 1 \cdot 5 = 1, \qquad -3 \cdot 3 + 2 \cdot 5 = 1.$$

To prove the theorem note first that neither a nor b can be zero unless the other is ± 1. In that case the theorem is trivial. Otherwise consider the set of all numbers of the form $ax + by$, where x and y are integers. If we choose $x = 1, y = 0$, and then $x = -1, y = 0$, it is clear that a and $-a$ are both in the set. Since $a \neq 0$, one of a and $-a$ is positive, so the set contains some positive numbers. Let d be the smallest positive number in the set, and write $d = as + bt$.

By Theorem 1.1 we can find q and r so that

$$b = dq + r, \qquad\qquad 0 \leq r < d.$$

Then

$$r = b - dq = b - (as + bt)q = a(-sq) + b(1 - qt),$$

so that r is also in the set. Now $0 < r < d$ is not possible, since d is the *least* positive number in the set. The only alternative is $r = 0$. Hence $b = dq$. A similar argument, beginning with

$$a = dq' + r', \qquad\qquad 0 \leq r' < d$$

shows that $r' = 0, a = dq'$.

This proves that d is a factor shared by both a and b. But a and b are relatively prime, so that $d = \pm 1$; moreover d is positive, so it must be 1. Hence $1 = as + bt$.

In what follows the notation "$m \mid n$" means "m divides n" or "m is a factor of n". If m is not a factor of n we write $m \nmid n$. The following theorem is the key to unique factorization.

THEOREM 1.3. *If p is a prime number and $p \mid ab$, then $p \mid a$ or $p \mid b$.*

The possibility that $p \mid a$ and $p \mid b$ is not excluded by the theorem.

If $p \mid a$ there is nothing to prove. Suppose then that $p \nmid a$; we shall show that in this case p must divide b.

Since p and a are relatively prime there exist integers l and m for which

$$lp + ma = 1, \qquad lpb + mab = b.$$

This follows from the preceding theorem. Since $p \mid ab$ we can write $ab = pq$. The last formula becomes $p(lb + mq) = b$, so that $p \mid b$.

COROLLARY 1.4. *If a prime number p divides a product $a_1 a_2 \cdots a_n$ of integers, it divides at least one of the a_i.*

For if p divides no a_i, then by Theorem 1.3 it cannot divide any of

$$a_1 a_2, \ (a_1 a_2) a_3, \ \ldots, \ (a_1 a_2 \cdots a_{n-1}) a_n.$$

We are now in a position to prove the fundamental theorem stated in the opening paragraph of the chapter. Let m be a positive integer not 1. If it is not prime suppose it factors as $m = m_1 m_2$, where $m_1 > 1$, $m_2 > 1$. If m_1 and m_2 are primes, stop; otherwise repeat the procedure for m_1 and m_2, and continue it for the new factors which appear. Eventually we must arrive at a stage where none of the factors will decompose again. Otherwise m, which is a finite integer, would be the product of an arbitrarily large number of factors all greater than 1.

Thus we arrive at a factorization

$$m = p_1 p_2 \cdots p_r,$$

where each p_i is positive and prime. Suppose

$$m = q_1 q_2 \cdots q_s$$

is any other factorization of m into positive primes. We must prove that the two factorizations differ at most in the order in which the primes appear. Since

$$p_1 p_2 \cdots p_r = q_1 q_2 \cdots q_s$$

it follows from Corollary 1.4 that q_1 must divide one of the p_i. We may suppose it to be p_1, by renumbering the p_i if necessary. Then $q_1 \mid p_1$. Since p_1 and q_1 are positive and prime $p_1 = q_1$. Hence, dividing out $p_1 = q_1$, we obtain

$$p_2 \cdots p_r = q_2 \cdots q_s.$$

This procedure can be repeated with $q_2, \ldots,$ until all the prime factors on one side are exhausted. At this stage all the factors on the other side must also be exhausted; otherwise we should have a product of primes on one side equal to 1 on the other. Then $r = s$ and we are done.

If we try to apply the principle of unique factorization to negative integers, we encounter an obvious difficulty in the possible presence of minus signs in the factors. Thus

$$-12 = 2^2(-3) = (-2)(-3)(-2)$$

are two factorizations of -12 into primes, and these factorizations differ not merely in the order of the factors, but in the factors themselves. For in the first case the factors are 2, 2, -3; in the second case they are -2, -3, -2. This difficulty can be remedied by a slight restatement of the fundamental theorem to include negative numbers. Let 1 and -1 be called *units*. The new statement is this.

THEOREM 1.5. (*The Fundamental Theorem*). *Each integer not zero or a unit can be factored into the product of primes which are uniquely determined to within order and multiplication by units.*

The slight change in the original proof which is needed here will be left to the reader.

2. **A general problem.** We are now in a position to state the basic problem of algebraic number theory: if we

extend the meaning of "integer" to include a wider class of numbers than the numbers $0, \pm 1, \pm 2, \ldots$ is there still a valid analogue of Theorem 1.5? The nature of the question can best be made plain by examples.

For this purpose we select first the *Gaussian* integers. By such an integer we shall mean a number of the form $a + bi$, where a and b are ordinary integers, and $i = \sqrt{-1}$. To avoid confusion later we shall refer to the ordinary integers as the *rational* integers. Let G denote the set of all Gaussian integers, and J the set of all rational integers. We shall sometimes write $\alpha \in A$ if α is a member of a set A. Note that in each set the sum, difference and product of integers are integers.

If α and β are numbers in G we say that α divides β, written $\alpha \mid \beta$, if there is a number γ in G such that $\beta = \alpha\gamma$. An element of G is a *unit* if it divides 1, and hence also every element of G. A number π is *prime* if it is not a unit and if in every factorization $\pi = \alpha\beta$ one of α or β is a unit. With this terminology Theorem 1.5 becomes meaningful for the integers of G.

But is it *true*? It is, as we shall show presently. This fact may strike the reader as only what is to be expected. That such an impression is erroneous we demonstrate by exhibiting another simple class of "integers" for which Theorem 1.5 is meaningful, but false.

Let us now mean by an "integer" any number of the form $a + b\sqrt{-5}$, where a and b are rational integers. Clearly the sum, difference and product of such integers are of the same form. We shall denote the totality of them by H. Define unit and prime just as we did for the Gaussian integers by simply reading H for G wherever the latter occurs. As we shall prove a little later, ± 1 are the only units in H; the numbers 3, 7, $1 + 2\sqrt{-5}$, $1 - 2\sqrt{-5}$

will turn out to be prime in H. But observe that

$$21 = 3 \cdot 7 = (1 + 2\sqrt{-5})(1 - 2\sqrt{-5}),$$

so that the factorization of 21 into prime factors is *not* unique to within order and multiplication by units.

It is therefore reasonable to ask for which classes of "integers" the fundamental theorem holds, and for which it does not. In particular how does one explain the discrepancy in behavior between the sets J and G on the one hand and H on the other? The answer to these questions must be postponed until later. For the present we content ourselves with demonstrating the assertions just made concerning the sets G and H.

3. **The Gaussian integers.** If $\alpha = a + bi$ is an element of G its *norm* $N(\alpha)$, or simply $N\alpha$, is defined to be $\alpha\bar{\alpha} = |\alpha|^2 = a^2 + b^2$. ($\bar{\alpha}$ is the complex-conjugate of α). The following list contains the fundamental properties of the norm.

   (i) If α is in J as well as in G, then $N\alpha = \alpha^2$.
   (ii) $N(\alpha\beta) = N\alpha N\beta$.
   (iii) $N\alpha = 1$ if and only if α is a unit.
   (iv)

$$N\alpha \begin{cases} = 0 & \text{if } \alpha = 0, \\ = 1 & \text{if } \alpha = \pm 1 \text{ or } \pm i, \\ > 1 & \text{otherwise.} \end{cases}$$

   (v) If $N\alpha$ is prime in J, then α is prime in G.
   The proof of (i) is obvious since $b = 0$. To prove (ii)

observe that if $\alpha = a + bi$, $\beta = c + di$, then

$$(\alpha\beta)\,(\overline{\alpha\beta}) = (\alpha\bar{\alpha})\,(\beta\bar{\beta}).$$

As for (iii), suppose first that α is a unit. Then $\alpha \mid 1$, so $\alpha\beta = 1$ for some β. By (ii) $N\alpha N\beta = N1 = 1$, and $N\alpha \mid 1$. Since $N\alpha$ must be a non-negative integer, $N\alpha = 1$. Conversely if $N\alpha = 1$, $a^2 + b^2 = 1$, so that $a = 0$ or $b = 0$. Then $\alpha = 1, -1, i$ or $-i$, and these are obviously units. This argument also establishes most of (iv); the rest we leave to the reader.

Finally to prove (v), suppose $N\alpha$ is prime and $\alpha = \beta\gamma$. Then $N\alpha = N\beta N\gamma$ is prime in J. So one of $N\beta$ or $N\gamma$ is equal to 1, and by (iii) either β or γ is a unit.

The converse of (v) is false. To see this it is enough to show that 3 is prime in G, for $N3 = 3^2 = 9$. Suppose $3 = \alpha\beta$. Then $9 = N\alpha N\beta$. If neither α nor β is a unit $N\alpha \neq 1$, $N\beta \neq 1$, so $N\alpha = N\beta = 3$. But this means that if $\alpha = a + bi$, then $a^2 + b^2 = 3$; this is impossible for any pair of integers a, b in J. (why?)

In proving that Theorem 1.5 holds for the Gaussian integers we shall imitate as far as possible the proof already given for rational integers.

THEOREM 1.6. *If α and β are Gaussian integers, $\beta \neq 0$, then there exist two integers π and ρ such that*

$$\alpha = \pi\beta + \rho, \qquad\qquad N\rho < N\beta.$$

Since $\alpha/\beta = \alpha\bar{\beta}/\beta\bar{\beta}$ it follows that $\alpha/\beta = A + Bi$, where A and B are ordinary rational numbers. Choose rational integers s and t such that

$$|A - s| \leq \tfrac{1}{2}, \qquad |B - t| \leq \tfrac{1}{2}.$$

This we can always do by choosing s and t as rational integers nearest to A and B respectively. Now let $\pi = s + ti$, $\rho = \alpha - \pi\beta$.

To show that $N\rho < N\beta$ observe that

$$|\rho| = |\alpha - \pi\beta| = |\alpha - (s + ti)\beta|$$

$$= |\beta| \left|\frac{\alpha}{\beta} - s - ti\right| = |\beta| |(A - s) + (B - t)i|$$

$$= |\beta| \{(A - s)^2 + (B - t)^2\}^{1/2}$$

$$\leq |\beta| \left\{\frac{1}{2^2} + \frac{1}{2^2}\right\}^{1/2} < |\beta|.$$

Since $N\rho = |\rho|^2 < |\beta|^2 = N\beta$, the inequality is established.

As an example let $\alpha = 5 - i$, $\beta = 1 + 2i$. Then

$$\frac{\alpha}{\beta} = \frac{(5 - i)(1 - 2i)}{(1 + 2i)(1 - 2i)} = \frac{3}{5} - \frac{11}{5} i,$$

so $A = \frac{3}{5}$, $B = -\frac{11}{5}$. Take $s = 1$, $t = -2$, $\pi = 1 - 2i$, $\rho = (5 - i) - (1 - 2i)(1 + 2i) = 5 - i - 5 = -i$. Then

$$5 - i = (1 - 2i)(1 + 2i) - i,$$

and $N(-i) < N(1 + 2i)$.

Let the reader show by an example that, in contrast to Theorem 1.1, π and ρ are *not* uniquely determined.

THEOREM 1.7. *If π is a prime and $\pi \mid \alpha\beta$, then $\pi \mid \alpha$ or $\pi \mid \beta$.*

If $\pi \mid \alpha$ we are done; so suppose $\pi \nmid \alpha$. We shall prove that $\pi \mid \beta$.

By Theorem 1.6 we can find δ and ρ so that

$$\alpha = \delta\pi + \rho, \qquad N\rho < N\pi.$$

Moreover $N\rho \neq 0$, for otherwise $\rho = 0$ so that $\pi \mid \alpha$, contrary to assumption. So $0 < N\rho < N\pi$.

Consider the set T of all non-zero integers of G having the form $\alpha\xi + \pi\eta$, where ξ and η are in G. $\rho = \alpha - \pi\delta$ is an integer in T. By property (iv) of norms in G, every element in T has norm at least equal to 1, so there must be one of them $\gamma = \alpha\xi_0 + \pi\eta_0$ which is of least positive norm. Now $\rho = \alpha - \pi\delta$ is in T and has norm less than $N\pi$. Since γ is of least norm, then also $N\gamma < N\pi$. We show next that γ is actually a unit.

Choose θ and ζ so that

$$\pi = \theta\gamma + \zeta, \qquad\qquad N\zeta < N\gamma.$$

Since $\zeta = \pi - \theta\gamma = \pi - \theta(\alpha\xi_0 + \pi\eta_0) = \alpha(-\theta\xi_0) + \pi(1 - \theta\eta_0)$, $N\zeta = 0$, for if $N\zeta \neq 0$, then ζ would be an element of T of smaller norm than γ. So $\zeta = 0$ and $\pi = \theta\gamma$, $N\pi = N\theta N\gamma$. One of θ and γ is a unit since π is a prime. But if $N\theta = 1$, then $N\pi = N\gamma$, which contradicts $N\pi > N\gamma$. So θ is not a unit, which means that γ is.

Hence $\gamma = \alpha\xi_0 + \pi\eta_0$ is a unit. Now

$$\alpha\beta\xi_0 + \pi\beta\eta_0 = \gamma\beta.$$

Since $\pi \mid \alpha\beta$ by hypothesis and $\pi \mid \pi\beta\eta_0$, then also $\pi \mid \gamma\beta$. So $\gamma\beta = \pi\tau$ for some τ in G. Then $\beta = \pi(\tau/\gamma)$ and $\pi \mid \beta$, for τ/γ is in G.

To prove that Theorem 1.5 is valid for the integers of G we proceed much as in the case of the rational integers. If α is not a unit or a prime let $\alpha = \alpha_1\alpha_2$, where $N\alpha_1 > 1$, $N\alpha_2 > 1$. Repeat this procedure for α_1 and α_2, and continue it. It must stop sometime, for otherwise $N\alpha$ would be the product of an arbitrarily large number of factors each greater than 1. So $\alpha = \pi_1 \cdots \pi_r$, where the π_i are primes. If also $\alpha = \sigma_1 \cdots \sigma_t$, where the σ_i are primes, then by Theorem 1.7 σ_1 must divide one of the π_i, say π_1. Hence $\sigma_1 = \pi_1\epsilon_1$, where ϵ_1 is a unit. Then

$$\pi_2 \cdots \pi_r = \epsilon_1\sigma_2 \cdots \sigma_t.$$

We can now complete the proof as we did for J.

It remains finally to establish the still unproved statements about H made in the preceding section, namely that ± 1 are the only units, and that 3, 7, $1 + 2\sqrt{-5}$, $1 - 2\sqrt{-5}$ are prime numbers in H.

If $\alpha = a + b\sqrt{-5}$, define $N\alpha = \alpha\bar{\alpha} = a^2 + 5b^2$. As before, $N(\alpha\beta) = N\alpha N\beta$. α is a unit if and only if $N\alpha = 1$; the proof goes as in the case of the Gaussian integers. But $a^2 + 5b^2 = 1$ only when $b = 0$, $a = \pm 1$, so $\alpha = \pm 1$ are the only units in H.

To show that 3 is a prime, suppose $3 = \alpha\beta$, where neither α nor β is a unit—that is, $N\alpha \neq 1$, $N\beta \neq 1$. Since $9 = N3 = N\alpha \cdot N\beta$, then $N\alpha = N\beta = 3$, so $a^2 + 5b^2 = 3$. If $b \neq 0$ then $a^2 + 5b^2 > 3$, so b must be zero. But then $a^2 = 3$, which cannot occur for an integer a in J. Similarly if $7 = \alpha\beta$, $N\alpha \neq 1$, $N\beta \neq 1$, then $a^2 + 5b^2 = 7$. If $b^2 \neq 0$, $b^2 \neq 1$ then $a^2 + 5b^2 > 7$. So either $b = 0$, $a^2 = 7$, which is impossible, or $b = \pm 1$, $a^2 = 2$, which is also impossible.

The numbers $1 \pm 2\sqrt{-5}$ are prime, for if $1 \pm 2\sqrt{-5} = \alpha\beta$, then $N(1 \pm 2\sqrt{-5}) = 21 = N\alpha N\beta$. Unless one of α or β is a unit $N\alpha = 3$ or $N\beta = 3$, and this possibility has already been excluded.

An additional example of a class of "integers" for which unique factorization is valid is given by the set of numbers $a + b\omega$, where $\omega = \frac{1}{2}(-1 + \sqrt{-3})$. The reader who is interested in the details will find them given in Chapter XII of the book of Hardy and Wright listed in the bibliography.

Problems

1. Determine whether 7 and 0 are relatively prime.
2. (i) Show that 123 and 152 are relatively prime.
   (ii) Find integers s and t such that $123s + 152t = 1$.

Hint: Theorem 1.2 gives no method for finding s and t, but by using Theorem 1.1 repeatedly you can obtain $152 = 1 \cdot 123 + 29$, $123 = 4 \cdot 29 + 7$, $29 = 4 \cdot 7 + 1$. Now eliminate the 7's and the 29's to express 1 in terms of 123 and 152. This method is called the *Euclidean algorithm*.

3. If a, b, c are rational integers such that $ab = ac$ and $a \neq 0$, show that $b = c$.

4. Establish these properties of division in G: if α, β, $\gamma \in G$, then
   (a) $\alpha \mid \beta \Rightarrow \alpha \mid \beta\gamma$;
   (b) $\alpha \mid \beta, \beta \mid \gamma \Rightarrow \alpha \mid \gamma$;
   (c) $\alpha \mid \beta, \alpha \mid \gamma \Rightarrow \alpha \mid (\beta + \gamma)$.
   The symbol "$\Rightarrow$" is to be read "implies".

5. Give another example of non-unique factorization in H. Make sure that the factors you obtain are actually prime in H.

6. Let $\alpha, \beta \in G$, and $\beta = b_1 + ib_2$, where b_1 and b_2 are real. Show that if $N\alpha \mid b_1$ and $N\alpha \mid b_2$ then $\alpha \mid \beta$ in G. Is the converse of this assertion true?

7. Determine which of the following are primes in G: $1 + 4i$, 7, $13i$, $1 + 3i$.

8. Find all primes in G among the numbers $a + bi$, where a and b are real and of absolute value at most 5. Time saving hint: show that except when $a = b = 1$ you may suppose $0 \leqq a < b$.

9. Factor each of these Gaussian integers into the product of Gaussian primes: $7 + 70i$, $1 - 7i$, $17 + i$, $219 + 219i$. Suggestion: factor the norm of each integer.

10. Find all non-unit common divisors of $9 + 3i$ and $-3 + 7i$.

11. Show that the norm of a Gaussian integer $a + bi$ cannot be of the form $4k + 3$.

12. Find *all* π and ρ in G such that

$$5 - i = \pi(1 + 2i) + \rho, \qquad N\rho < N(1 + 2i).$$

13. The number π in the expression $\alpha = \pi\beta + \rho$ occurring in Theorem 1.6 need not be unique. Find the number of values of π which can occur, depending on the position of α/β in the complex plane. Remember that $N\rho < N\beta$.

14. Find an analogue for Theorem 1.2 in G.

15. Using the technique suggested in Problem 2, find ξ, η in G such that

$$(3 + 13i)\xi + (2 + 5i)\eta = 1.$$

16. Show that division theorems like Theorem 1.1 and Theorem 1.6 are not valid in H by showing that there exist no π and ρ in H for which $1 + \sqrt{-5} = 2\pi + \rho$, and $N\rho < N2$.

17. Let π_1 and π_2 be primes in G such that $\pi_1 \nmid \pi_2$. If $\alpha \in G$, and $\pi_1 \mid \alpha$ and $\pi_2 \mid \alpha$ show that $\pi_1\pi_2 \mid \alpha$.

18. Show that the norm in H has these properties:
    (a) $N(\alpha\beta) = N(\alpha)N(\beta)$;
    (b) $N\alpha = 1$ if and only if α is a unit in H;
    (c) $N\alpha = 0$ if and only if $\alpha = 0$;
    (d) if $N(\alpha)$ is prime in J, then α is prime in H. Show also that $3 + 2\sqrt{-5}$ and $2 + \sqrt{-5}$ are primes in H.

CHAPTER II

THE GAUSSIAN PRIMES

1. Rational and Gaussian primes. It is not difficult
to establish the existence of an infinite number of rational
primes—that is, primes in J. The simplest proof, due to
Euclid, goes as follows. Suppose p_1, p_2, ..., p_n are known
to be prime. Then the number $N = 1 + p_1 p_2 \cdots p_n$
cannot have any one of the p_i as a factor, since then 1
would also have that p_i as a factor. Then any prime factor
of N is different from p_1, ..., p_n. This means that if
any finite set of prime numbers is given, there is a prime
different from any of them; so there are an infinite number
if there is at least one. But 2 is a prime, and the conclusion
follows.

Precisely the same proof is valid for Gaussian primes
provided only that we can find one prime. But 3 has
already been shown to be a Gaussian prime, so that G
contains an infinity of primes. We can accomplish con-
siderably more: we shall characterize explicitly all the
primes in G in terms of those in J. In order to achieve this
we shall need some material from elementary number
theory. Actually we shall prove somewhat more than we
need for the present purpose. The additional results will
find application later.

2. Congruences. In this section we deal only with
rational integers.

Let m be an integer not zero. Two integers a and b are said to be *congruent modulo m*, written

$$a \equiv b \pmod{m} \qquad or \qquad a \equiv b \ (m),$$

if $m \mid (a - b)$. If a and b are not congruent mod m we write $a \not\equiv b \ (m)$.

According to Theorem 1.1 every integer a leaves a remainder r, $0 \le r < |m|$, on division by $|m|$. We shall show that a and b are congruent modulo m if and only if they leave the same remainder on division by $|m|$. First suppose

$$a = q|m| + r, \qquad b = q'|m| + r, \qquad 0 \le r < |m|.$$

Then

$$a - b = (q - q')|m|, \qquad \pm m \mid (a - b),$$

so that $m \mid (a - b)$. Conversely suppose $a \equiv b \ (m)$. Let $a = q|m| + r$, $b = q'|m| + r'$, $0 \le r < |m|$, $0 \le r' < |m|$. Then

$$a - b = (q - q')|m| + (r - r').$$

Since $|m|$ divides $a - b$, $|m|$ divides $r - r'$. But $-|m| < r - r' < |m|$, so $r - r'$ cannot be divisible by $|m|$ unless $r = r'$.

The following properties of congruences will be used frequently.

(i) If $a \equiv b \ (m)$, then $b \equiv a \ (m)$.

(ii) If $a \equiv b \ (m)$ and $b \equiv c \ (m)$, then $a \equiv c \ (m)$.

(iii) If $a \equiv b \ (m)$, then $ka \equiv kb \ (m)$ for any integer k.

(iv) If $a_i \equiv b_i \ (m)$ for $i = 1, 2, \ldots, n$, then

$$a_1 + a_2 + \cdots + a_n \equiv b_1 + b_2 + \cdots + b_n \, (m),$$

$$a_1 a_2 \cdots a_n \equiv b_1 b_2 \cdots b_n \, (m).$$

The last part of (iv) is the only one of these properties which is not quite obvious. We verify it when $n = 2$; the general case follows by repeated application of this one. By (iii)

$$a_1a_2 \equiv b_1a_2\,(m), \qquad b_1a_2 \equiv b_1b_2\,(m),$$

so that by (ii), $a_1a_2 \equiv b_1b_2\,(m)$.

It is not true that if $ka \equiv kb\,(m)$, then $a \equiv b\,(m)$. For example $3\cdot 2 \equiv 3\cdot 1\,(3)$, but $2 \not\equiv 1\,(3)$. In order to state a correct converse of (iii) we introduce the notion of the *greatest common divisor* (h, k) of two integers h and k; it is simply the largest positive factor common to both h and k. Note that if c is *any* common factor of h and k, then $c \mid (h, k)$; this follows from the fundamental theorem of arithmetic. We can now state

(v) If $ka \equiv kb\,(m)$, then $a \equiv b\,(\mathrm{mod}\ m/d)$, where $d = (k, m)$. In particular, $a \equiv b\,(\mathrm{mod}\ m)$ if k and m are relatively prime, that is $d = 1$.

Now suppose m to be a positive integer. Since *every* integer leaves on division by m one of the remainders $0, 1, \ldots, m - 1$, every integer is congruent to exactly one of these integers modulo m. Any set of integers such that every integer is congruent to exactly one of them modulo m is called a *complete residue* (or *remainder*) *system modulo* m. It follows that a set of integers is a complete residue system modulo m if and only if it consists of exactly m integers, no two of which are congruent modulo m.

THEOREM 2.1. *If $a_1, a_2, \ldots, a_m$ form a complete residue system modulo m, and if $(a, m) = 1$, then $aa_1, aa_2, \ldots, aa_m$ also form such a system.*

For if $aa_i \equiv aa_j\,(m)$, then $a_i \equiv a_j\,(m)$, by property (v) above.

THEOREM 2.2. *(Fermat). If p is a prime and $(a, p) = 1$, then $a^{p-1} \equiv 1\,(p)$.*

The numbers $0, 1, 2, \ldots, p - 1$ form a complete residue system modulo p. Hence $0, a, 2a, \ldots, (p - 1)a$ do also, by the preceding theorem. Now each number on one list is congruent to exactly one on the other. Omitting 0 from each list, since the zeros correspond, we get by (iv)

$$a \cdot 2a \, \cdots \, (p - 1)a \equiv 1 \cdot 2 \, \cdots \, (p - 1) \quad (\bmod\ p),$$

or

$$(p - 1)! a^{p-1} \equiv (p - 1)! \quad (\bmod\ p).$$

By (v) we can divide out $(p - 1)!$ from each side to obtain the conclusion.

COROLLARY 2.3. *If p is a prime, then $a^p \equiv a(p)$, for any integer a.*

THEOREM 2.4. (*Wilson*). *If p is a prime, then $(p - 1)! \equiv -1(p)$.*

If $p = 2$ or $p = 3$ the conclusion is obvious, so suppose $p > 3$.

Let a be one of the numbers $1, 2, \ldots, p - 1$, and let us examine the equation $ax \equiv 1(p)$. Note that $(a, p) = 1$. If x goes through the values $1, 2, \ldots, p - 1$ then by Theorem 2.1 ax goes through a complete residue system mod p, excepting 0. Hence there is one and only one x which satisfies the congruence.

Then the numbers $1, 2, \ldots, p - 1$ fall into pairs such that the product of any pair is congruent to 1 modulo p. If the members of a pair are equal, say to a, then $a^2 \equiv 1$, $a^2 - 1 \equiv 0$, $p \mid (a - 1)(a + 1)$, so $p \mid (a + 1)$ or $p \mid (a - 1)$. p cannot divide both $a + 1$ and $a - 1$, since it would divide their difference 2. Hence $a \equiv 1(p)$ or $a \equiv -1(p)$. Since $1 \leq a \leq p - 1$ we have that either $a = 1$ or $a = p - 1$.

With the $p - 3$ numbers of the set $2, \ldots, p - 2$ we

can form the product of the $(p-3)/2$ pairs to obtain

$$2\cdot3\cdot4\cdots(p-2) \equiv 1 \qquad (p).$$

Then $(p-1)! \equiv p-1 \equiv -1(p)$.

COROLLARY 2.5. *If p is a prime number of the form $4m+1$, then $p \mid (n^2+1)$, where $n = (2m)!$*

Consider the two sets of numbers

$$-1, -2, \ldots, -2m$$

$$4m, 4m-1, \ldots, 2m+1.$$

Each element of the lower row is congruent modulo p to the element of the upper row directly above, since their difference is p. Hence

$$4m(4m-1)\cdots(2m+1) \equiv (-1)(-2)\cdots(-2m) \qquad (p).$$

Since also $(2m)! \equiv (2m)!$, multiplication yields

$$(4m)! \equiv \{(2m)!\}^2 \qquad (p).$$

Let $n = (2m)!$. Since $(4m)! = (p-1)! \equiv -1$ by Wilson's theorem, it follows that $-1 \equiv n^2(p)$.

THEOREM 2.6. *If p is a prime and a and b are integers, then*

$$a^p + b^p \equiv (a+b)^p \qquad (mod\ p).$$

By Corollary 2.3, $c^p \equiv c(p)$ for any integer c. Let $c = a + b$. Then $(a+b)^p \equiv a + b$. But also $a^p \equiv a$, $b^p \equiv b$, and from these the result follows.

3. **Determination of the Gaussian primes.** We are now in a position to classify the Gaussian primes. The situation is somewhat complicated by the fact that a rational prime can cease to be a prime in G—for example, $5 = (1+2i)(1-2i)$; part of our problem is to decide which rational primes are also Gaussian primes.

It is convenient in the classification to call two Gaussian integers *associates*, written $\alpha \sim \beta$, if $\alpha \mid \beta$ and $\beta \mid \alpha$—that is, if $\alpha = \beta\epsilon$ where ϵ is a unit.

THEOREM 2.7. *The Gaussian primes fall into the following three classes*:

1. *all positive rational primes of the form* $4m + 3$ *and their associates in* G;
2. *the number* $1 + i$ *and its associates*;
3. *all integers associated with either* $x + iy$ *or* $x - iy$ *where* $x > 0$, $y > 0$, x *is even, and* $x^2 + y^2$ *is a rational prime of the form* $4m + 1$.

Before proving the theorem we illustrate its application in detecting Gaussian primes. Let $p = 3$. This is in the first of the classes mentioned in the theorem, with $m = 0$; hence 3 is a Gaussian prime. Let $p = 5$. This is of the form $4m + 1$, and $5 = (2 + i)(2 - i)$, so $2 + i$, $2 - i$ and their associates are primes, by the third part of the theorem.

To prove the theorem we show first that any prime π in G divides exactly one positive rational prime p. For $N\pi = \pi\bar{\pi}$, so $\pi \mid N\pi$. Let $N\pi = p_1 \cdots p_r$ be the decomposition in J of $N\pi$ into positive primes. Then $\pi \mid p_1 \cdots p_r$. By Theorem 1.7 π divides one of the p_i. So π divides some positive rational prime. It cannot divide two, p and q. For by Theorem 1.2 we can find rational integers l and m such that $lp + mq = 1$. If $\pi \mid p$, $\pi \mid q$ then $\pi \mid 1$, so π is a unit, not a prime, contrary to hypothesis.

Hence we can get each prime in G once and only once by considering the factorization of all positive rational primes, treated as elements of G.

Now let π be a prime, and p the positive prime for which $\pi \mid p$. Then $N\pi \mid Np$. But $Np = p^2$, since p is a rational integer. Hence $N\pi = p$ or $N\pi = p^2$. If $\pi = x + iy$ then $x^2 + y^2 = p$ or $x^2 + y^2 = p^2$.

Divide p by 4. According to Theorem 1.1 this leaves a remainder of 1, 2 or 3. We consider the three cases separately.

Case 1. $p \equiv 3\,(4)$. As stated just above, $x^2 + y^2 = p$ or $x^2 + y^2 = p^2$. It will be shown now that the first of these two possibilities cannot occur. Since p is odd, one of x and y, say x, must be even, the other odd; otherwise the sum of their squares would be even. Let $x = 2a$, $y = 2b + 1$. Then if $x^2 + y^2 = p$,

$$p = x^2 + y^2 = 4a^2 + (2b + 1)^2$$

$$= 4(a^2 + b^2 + b) + 1 \equiv 1\,(4),$$

whereas $p \equiv 3$.

So in this case $x^2 + y^2 = p^2$, and $N\pi = Np$. Since $\pi \mid p$, $p = \pi\gamma$, where γ is in G. Then $Np = N\pi N\gamma$, $N\gamma = 1$, γ is a unit, and $p \sim \pi$.

This accounts for the first part of Theorem 2.7.

Case 2. $p \equiv 2\,(4)$. In this case $p = 2$, since this is the only even prime. But $2 = (1 + i)(1 - i)$, and $\pi \mid 2$. So $\pi \mid (1 + i)$ or $\pi \mid (1 - i)$. But $N(1 + i) = N(1 - i) = 2$, a rational prime. We showed earlier that if $N\alpha$ is prime so is α. Then $1 + i$ and $1 - i$ are prime. Hence $\pi \sim 1 + i$ or $\pi \sim 1 - i$. Since $(1 + i)/(1 - i) = i$, $1 + i \sim 1 - i$, and the second part of the theorem is done.

Case 3. $p \equiv 1\,(4)$. p is of the form $1 + 4m$, so that Corollary 2.5 is applicable and $p \mid n^2 + 1$ for some rational integer n. But $n^2 + 1 = (n + i)(n - i)$ and $\pi \mid p$, so $\pi \mid (n + i)$ or $\pi \mid (n - i)$. But p does not divide $n + i$ or $n - i$, for otherwise one of $(n \pm i)/p$ would be a Gaussian integer; this cannot be, for $1/p$ is not a rational integer. Hence π and p are not associated. It follows that

$N\pi \neq Np$, so $x^2 + y^2 \neq p^2$. From our earlier remarks, this leaves only the alternative $x^2 + y^2 = p$.

Then $\pi\bar{\pi} = p$. Now $\pi = x + iy$ is prime by assumption; so is $\bar{\pi} = x - iy$, since $N\bar{\pi} = p$. They are not associated, for otherwise $x + iy = \epsilon(x - iy)$, where $\epsilon = 1, -1, i$ or $-i$. If $\epsilon = 1$, $y = 0$, $x^2 = p$, so p is not a prime. If $\epsilon = -1$, $x = 0$, $y^2 = p$, and the same conclusion follows. If $\epsilon = \pm i$, $x = \pm y$ and p is even. All of these eventualities are impossible, so $x + iy$ and $x - iy$ are not associated.

Finally, since $x^2 + y^2 = p$, one of x and y must be even, the other odd. In order to give an associated prime having positive even real part, it may be necessary to multiply π by $\pm i$ or -1.

4. Fermat's theorem for Gaussian primes.

It is now reasonable to ask whether the theory discussed in §2 for rational primes has an analogue for Gaussian integers. This is the case, and the theory of congruences and complete residue systems can be carried over. Since we expect to investigate these things later for far more general classes of numbers than the Gaussian integers, we shall only illustrate the kind of thing to be expected by proving the analogue of Fermat's Theorem 2.2.

If α, β, γ are in G then by $\alpha \equiv \beta \pmod{\gamma}$ or $\alpha \equiv \beta (\gamma)$ we shall mean that $\gamma \mid (\alpha - \beta)$ in G. We say that α and β are *relatively prime* if they have no common factors in G except units.

THEOREM 2.8. (*Analogue of Fermat's theorem*). *If π is a prime in G and α an element of G relatively prime to π, then*

$$\alpha^{N\pi-1} \equiv 1 (\pi).$$

Let p be the unique positive prime p, discussed in the proof of the preceding theorem, for which $\pi \mid p$. There are three cases, corresponding to the three parts of Theorem 2.7.

Case 1. $p \equiv 3\,(4)$. In this case $N\pi = x^2 + y^2 = p^2$, so we must show $\alpha^{p^2-1} \equiv 1\,(\pi)$. What we shall prove is that $\alpha^{p^2} \equiv \alpha\,(p)$. From this the result will follow, for

$$\pi \mid p, \qquad p \mid (\alpha^{p^2} - \alpha), \qquad \pi \mid \alpha(\alpha^{p^2-1} - 1)$$

so $\pi \mid (\alpha^{p^2-1} - 1)$, since $\pi \nmid \alpha$.

Let $\alpha = l + im$. Then $\alpha^p \equiv l^p + i^p m^p\,(p)$, by the expansion of $(l + im)^p$ by the binomial expansion, and the observation that p divides all binomial coefficients except the first and last. Since p is of the form $4n + 3$, $i^p = -i$. Also $l^p \equiv l$, $m^p \equiv m$ by Corollary 2.3, so

$$\alpha^p \equiv l - im \equiv \bar{\alpha}\,(p).$$

Conjugation yields
$$\bar{\alpha}^p \equiv \alpha\,(p),$$
so that
$$\alpha^{p^2} \equiv \bar{\alpha}^p \equiv \alpha\,(p),$$
as asserted.

Case 2. $p \equiv 2\,(4)$. In this case $p = 2$, so that $\pi \sim 1 + i$. We may assume $\pi = 1 + i$. Since $N\pi = 2$, what we must prove is that $\alpha^{N\pi-1} = \alpha \equiv 1\,(\pi)$, or simply that $1 + i$ divides $\alpha - 1$ when $1 + i$ and α are relatively prime. Since $1 + i \nmid \alpha$, the division algorithm gives

$$\alpha = q(1 + i) + \rho, \qquad 0 < N\rho < N(1 + i) = 2.$$

Therefore, $N(\rho) = 1$, so $\rho = 1, -1, i$ or $-i$. It follows that $1 + i \mid \rho - 1$. Since $\alpha - 1 = q(1 + i) + (\rho - 1)$ we can conclude that $1 + i \mid \alpha - 1$.

Case 3. $p \equiv 1\,(4)$. Now $N\pi = x^2 + y^2 = p$, so we must show that $\alpha^{p-1} \equiv 1\,(\pi)$. Since $\pi \mid p$ and α, π are relatively prime this will follow if we can prove that $\alpha^p \equiv \alpha\,(p)$.

Let $\alpha = l + mi$. As in Case 1, $\alpha^p \equiv l^p + i^p m^p\,(p)$.

But p is of the form $4n + 1$, so that $i^p = i$ and $\alpha^p \equiv l + im = \alpha$, as required.

From the theorem we obtain the following consequence:

COROLLARY 2.9. *Let π be a prime and $\pi \nmid \alpha$. Then the congruence $\alpha x \equiv \beta(\pi)$ has solutions $x \equiv \alpha^{N\pi-2}\beta(\pi)$.*

Problems

1. Let $a_1, a_2, \ldots, a_m$ form a complete residue system mod m, and suppose b is an integer for which ba_1, $ba_2, \ldots, ba_m$ also form a complete residue system. Prove that $(b, m) = 1$.
2. Let $m \in J, m > 1$. Suppose that $(m - 1)! \equiv -1(m)$. Prove that m is prime.
3. Establish Theorem 2.6 by using the binomial expansion of $(a + b)^p$.
4. Show that if $a + bi$ is prime in G, so is $b + ai$.
5. Let b and c be relatively prime rational integers. Suppose $\alpha \mid b$, $\alpha \mid c$ where $\alpha \in G$. What can you conclude about α?
6. Let $\pi = a + bi, a \neq 0, b \neq 0$, be a prime in G. Show that if π is a factor of its conjugate $\bar{\pi}$, then π is an associate of $1 + i$.
7. Let π be a nonreal prime in G, distinct from $1 + i$ or its associates. Suppose $\alpha \in G$ and $\pi \mid \alpha$, $\pi \mid \bar{\alpha}$. Show that $N\pi \mid \alpha$.
8. Prove that every rational prime of the form $4m + 1$ can be expressed as a sum of two squares. Let p and q each be a rational prime of the form $4m + 1$. Prove that pq can be represented as a sum of two squares.
9. Show that there exist in G two elements distinct from units which are relatively prime but whose norms are not relatively prime.
10. If two elements of G have norms which are relatively

prime in J, then the elements are relatively prime in G.

11. Show that $-1 + 2i \mid 7^{16} - 1$.

12. Find a solution $x \in G$ of the congruence

$$(3 - 2i)x \equiv 1 \qquad (\bmod\ 1 - 2i)$$

by using Corollary 2.9. Check that the x you found is indeed a solution. Find all solutions of the congruence in G.

13. Find $\alpha_1, \ldots, \alpha_5$ in G such that $\alpha_j \not\equiv \alpha_k \pmod{2 + i}$ if $j \neq k$. Draw a diagram showing all points $a + bi \in G$ with $|a| \leq 4$, $|b| \leq 4$. Using five distinct symbols, show which points $a + bi$ are congruent to $\alpha_1, \ldots,$ to α_5. The five classes of integers constitute a complete residue system modulo $2 + i$. Can you prove this?

CHAPTER III

POLYNOMIALS OVER A FIELD

1. The ring of polynomials. A non-empty set S is called a *commutative ring* if there are two operations, denoted by $+$ and $\cdot$ such that for all a, b, $c \in S$, (i) $a + b = b + a \in S$, (ii) $(a + b) + c = a + (b + c)$, (iii) there is a 0 in S such that $a + 0 = a$ for all a in S, (iv) for each a in S there is an element $-a$ in S such that $a + (-a) = 0$. Further, (v) $a \cdot b = b \cdot a \in S$, (vi) $a \cdot (b \cdot c) = (a \cdot b) \cdot c$, and (vii) $a \cdot (b + c) = a \cdot b + a \cdot c$. A set satisfying (i)–(iv) is said to be a *commutative group under the operation* $+$. For example, the sets J, G, and H are commutative rings, as are the set of all rational numbers, which we denote by R, and the set of all complex numbers, which we denote by C. Note that the last two examples have a further property: they admit division by non-zero elements.

By a *number field* F we shall mean a commutative ring which contains J, is contained in C, and whose non-zero elements form a group under multiplication. Specifically, besides (v) and (vi), we have (viii) there is an element 1 different from 0 such that $a \cdot 1 = a$ for all a in F, and (ix) if $a \in F$ and $a \neq 0$ there is an element a^{-1} in F such that $a \cdot a^{-1} = 1$. Since ± 1, $\pm 2, \ldots \in F$, thus $\pm 1^{-1}$, $\pm 2^{-1}$, $\ldots \in F$ and hence all fractions of the form $mn^{-1} = m/n$, where $m, n \in J$ and $n \neq 0$, i.e., *any number field contains* R. R itself is a field. Further examples of number fields include the set of all real numbers, the set $\{a + b\sqrt{2} : a, b \in R\}$, and C. The sets J, G, and H are not number fields, for they do not contain R.

25

In abstract algebra one defines fields of a more general kind; in the present book, however, a "field" will always mean a "number field."

A *polynomial of degree n, $n \geq 0$*, over a field F is an expression of the form

$$p(x) = a_0 + a_1 x + \cdots a_{n-1} x^{n-1} + a_n x^n$$

where the coefficients a_i are in F and $a_n \neq 0$.

In what follows it is convenient to include 0 as a polynomial, but we give it no degree. The notation $f(x) \equiv 0$ will mean that $f(x)$ is the polynomial zero. A constant *not* zero satisfies the above definition of a polynomial of degree n, with $n = 0$.

We shall denote the collection of all polynomials by $F[x]$. We can add two polynomials in $F[x]$ or multiply a polynomial by a number from F. In both cases the result is another polynomial. If

$$q(x) = b_0 + b_1 x + \cdots + b_m x^m \in F[x],$$

we can form the product of $p(x)$ and $q(x)$ according to the rule

$$p(x)q(x) = c_0 + c_1 x + \cdots + c_k x^k,$$

where

$$c_0 = a_0 b_0$$

$$c_1 = a_0 b_1 + a_1 b_0$$

$$\cdots$$

$$c_i = a_0 b_i + a_1 b_{i-1} + \cdots + a_{i-1} b_1 + a_i b_0$$

and $k = m + n$. It is easy to see that $p(x)q(x)$ is in $F[x]$. We say that $p(x)$ divides $q(x)$ (in $F[x]$) and write $p(x) \mid q(x)$ if there exists an $r(x)$ in $F[x]$ such that $q(x) = p(x)r(x)$. The set of polynomials with coefficients in J is denoted by $J[x]$. We remark that all the concepts we have defined for $F[x]$ are meaningful for $J[x]$.

It is easy to verify that $F[x]$ forms a commutative ring and has a 1 element. However, $F[x]$ is not a field. (Why?) We shall see that $F[x]$ exhibits some of the same properties as J. This is reasonable in view of the similarity of their algebraic structure.

Each polynomial $p(x)$ of degree $n \geq 1$ can be factored uniquely into the form

$$p(x) = a_n(x - r_1)(x - r_2) \cdots (x - r_n).$$

Here the r_i are complex numbers which need not belong to the field F containing the coefficients of $p(x)$. This result is called the *fundamental theorem of algebra*. Its proof, however, depends upon methods from analysis, and we shall not give it here. As an example, $p(x) = x^2 + 2x + 3$ is a polynomial over the field R of rational numbers, but in this case $r_1 = -1 + \sqrt{-2}$, $r_2 = -1 - \sqrt{-2}$, and these are certainly not numbers in R.

The numbers $r_1, \ldots, r_n$ are called the *roots* or *zeros* of the polynomial. It follows from the unique factorization just mentioned that a polynomial of degree $n \geq 1$ has at most n distinct roots. It is of course possible for several or all of the roots to be identical. For example

$$x^3 - 3x^2 + 3x - 1 = (x - 1)(x - 1)(x - 1).$$

A polynomial over F is said to be *prime* or *irreducible* over F if it cannot be factored into a product of two or more polynomials

$$p(x) = p_1(x)p_2(x) \cdots p_k(x),$$

where each $p_i(x)$ is of lower degree than $p(x)$ and is itself a polynomial over F. For example $x^2 + 2x + 3$ is irreducible over R, although it is reducible over the field of all complex numbers.

We shall prove that every polynomial over F can be factored into the product of irreducible factors over F, and that the factorization is unique to within order and units. A *unit* is in this case simply a constant—that is, a number from F. Polynomials are relatively prime if they have only units as common factors.

The proof is not unlike that of the fundamental theorem of arithmetic, and we begin by establishing results which parallel the early theorems of Chapter I.

LEMMA 3.1. *Let $f(x)$ and $g(x)$ be polynomials of degrees n and m respectively over a field F, and suppose $n \geq m$. Then for a suitable number c in F the expression*

$$f(x) - cx^{n-m}g(x)$$

is identically zero or is a polynomial of degree less than n.

Let $f(x)$ and $g(x)$ be defined respectively by

$$f(x) = a_n x^n + a_{n-1} x^{n-1} + \cdots + a_0$$
$$g(x) = b_m x^m + b_{m-1} x^{m-1} + \cdots + b_0,$$

where $a_n \neq 0$, $b_m \neq 0$. Define $c = a_n/b_m$. Then

$$f(x) - cx^{n-m}g(x) = (a_n x^n + \cdots) - \frac{a_n}{b_m} x^{n-m}(b_m x^m + \cdots),$$

so that the term in x^n cancels. It is possible for all the terms to cancel, but in any case only terms of lower degree than x^n can survive.

THEOREM 3.2. *Let $f(x)$ and $g(x) \not\equiv 0$ be polynomials over F. Then there are polynomials $q(x)$ and $r(x)$ over F such that*

$$f(x) = q(x)g(x) + r(x),$$

where $r(x) \equiv 0$ or $r(x)$ is of lower degree than $g(x)$.

If $f(x)$ is identically zero or of lower degree than $g(x)$ we can take $q(x) \equiv 0$, and $r(x)$ to be $f(x)$ itself.

Now regard $g(x)$ as fixed, of degree m. We shall prove the theorem for all $f(x)$ of degree $n \geq m$ by induction. Suppose the conclusion of the theorem to be true for all $f(x)$ of degree between 0 and $n - 1$ inclusive. By the lemma

$$f(x) - cx^{n-m}g(x) = f_1(x)$$

is identically zero or of degree at most $n - 1$. By the first part of the proof if $f_1(x) \equiv 0$, or by the induction hypothesis if $f_1(x) \not\equiv 0$, we have

$$f_1(x) = q_1(x)g(x) + r(x),$$

where $r(x) \equiv 0$ or $r(x)$ is of lower degree than $g(x)$. Then

$$\begin{aligned} f(x) &= f_1(x) + cx^{n-m}g(x) \\ &= [cx^{n-m} + q_1(x)]g(x) + r(x) \\ &= q(x)g(x) + r(x), \end{aligned}$$

and the induction is complete.

THEOREM 3.3. *Let $f(x)$ and $g(x)$ be non-zero polynomials over F, relatively prime over F. Then there exist polynomials $s_0(x)$ and $t_0(x)$ over F such that*

$$1 = s_0(x)f(x) + t_0(x)g(x).$$

Consider the set T of all polynomials of the form $s(x)f(x) + t(x)g(x) \not\equiv 0$, where $s(x)$ and $t(x)$ have coefficients in F. Choose in T a member $d(x)$ of lowest degree. $d(x)$ may, of course, be a constant not zero. We shall show that it actually is.

By Theorem 3.2 we can find $q(x)$, $r(x)$ so that

$$r(x) = f(x) - q(x)d(x),$$

where $r(x) \equiv 0$ or $r(x)$ is of degree less than that of $d(x)$. The second of these possibilities is excluded, for $r(x)$ is obviously in T, and no polynomial in T is of lower degree than $d(x)$. So $r(x) \equiv 0$. Hence $f(x) = q(x)d(x)$. Similarly $g(x) = q_1(x)d(x)$ for some polynomial $q_1(x)$. Since $f(x)$ and $g(x)$ are relatively prime, $d(x)$ must be a constant $d \neq 0$. Since d is in T it has a representation

$$d = s_0(x)f(x) + t_0(x)g(x).$$

Divide by d, and the theorem is established.

A polynomial is *monic* if its leading coefficient a_n is 1. By use of Theorem 3.3 it is easy to prove the following two theorems which are analogous respectively to Theorems 1.3 and 1.5. The reader will find it a useful exercise to supply the details of the proofs.

THEOREM 3.4. *If $p(x)$, $f(x)$, $g(x)$ are polynomials over F, $p(x)$ irreducible, and $p(x)$ divides $f(x)g(x)$ over F, then $p(x)$ divides either $f(x)$ or $g(x)$.*

THEOREM 3.5. *Any polynomial $p(x) = a_n x^n + \cdots + a_0$ over F not zero or a constant can be factored into a product*

$$p(x) = a_n p_1(x) \cdots p_r(x)$$

where the $p_i(x)$ are irreducible monic polynomials over F, determined uniquely except for order.

2. The Eisenstein irreducibility criterion.

In this section we shall present a simple and useful test for the irreducibility of a polynomial over the field R of rational numbers.

A polynomial with rational integers as coefficients is *primitive* if the coefficients have no factors other than ± 1 common to all of them. In the other case, $f(x) \in J[x]$ is called *imprimitive*. Obviously the product of an imprimitive polynomial and an arbitrary polynomial in $J[x]$ is imprimitive. Here we shall give the converse proposition.

THEOREM 3.6. (*Gauss' Lemma*). *The product of primitive polynomials is primitive.*

Let $a_0 + a_1x + \cdots + a_nx^n$ and $b_0 + b_1x + \cdots + b_mx^m$ be primitive, and let $c_0 + c_1x + \cdots + c_kx^k$ be their product. Assume the product is not primitive. Then all the c_i are divisible by some prime number p. Let a_i and b_j be the *first* coefficients in the two original polynomials (note the order in which the terms were written) which are *not* divisible by p. They must exist, for the polynomials are primitive, and so not all their coefficients can be divisible by p.

Now, by the formula for the product of two polynomials,

$$c_{i+j} = (a_0b_{i+j} + \cdots + a_{i-1}b_{j+1}) + a_ib_j$$
$$+ (a_{i+1}b_{j-1} + \cdots + a_{i+j}b_0).$$

But $a_0, a_1, \ldots, a_{i-1}, b_0, b_1, \ldots, b_{j-1}$, and c_{i+j} are all divisible by p. So a_ib_j must also be divisible by p.

Since p is prime, $p \mid a_i$ or $p \mid b_j$. But this contradicts the choice of a_i and b_j as coefficients not divisible by p. Thus the assumption that the c_i have a factor p in common is erroneous, and $c_0 + c_1x + \cdots + c_kx^k$ must be primitive.

As an example, consider the primitive polynomials $x^2 + 3$ and $3x^2 + 7x - 11$. Their product

$$3x^4 + 7x^3 - 2x^2 + 21x - 33$$

is certainly primitive.

THEOREM 3.7. *If a polynomial with rational integral coefficients can be factored over R, it can be factored into polynomials with rational integral coefficients.*

For example,

$$2x^2 + 19x + 35 = (2x + 14)(x + \tfrac{5}{2}),$$

but also

$$2x^2 + 19x + 35 = (x + 7)(2x + 5).$$

The proof goes in two parts. First note that *any poly-nomial* $f(x) \not\equiv 0$ *over* R *can be written uniquely in the form*

$$f(x) = c_f f^*(x),$$

where $f^*(x)$ *is primitive and* c_f *is a positive rational number.* For suppose that

$$f(x) = a_n x^n + a_{n-1} x^{n-1} + \cdots + a_0,$$

where the a_i are rational numbers. We can write $a_i = b_i/c$, where c is the least positive common denominator of all the fractions a_i. Then

$$f(x) = \frac{1}{c}(b_n x^n + b_{n-1} x^{n-1} + \cdots b_0).$$

Now factor out of the expression in parentheses the largest positive factor common to all the b_i. Then what remains inside the parentheses we call $f^*(x)$, what is outside c_f. Clearly $c_f > 0$, and $f^*(x)$ is primitive by the very manner in which it is defined. As to the uniqueness, if

$$f(x) = c_f f^*(x) = cp(x),$$

where c_f and c are positive and $f^*(x)$, $p(x)$ are primitive, then $p(x) = (c_f/c)f^*(x) = (\alpha/\beta)f^*(x)$, where α and β are relatively prime positive integers. Since $p(x)$ has rational integer coefficients, $\beta \mid \alpha d_i$ for $0 \leq i \leq n$, where $d_0, \ldots, d_n$ are the coefficients of f^*. Since $(\alpha, \beta) = 1$, β divides all the d_i. Now $\beta = 1$, because f^* is primitive. Similarly, $\alpha = 1$. Thus $c_f = c$ and $f^*(x) = p(x)$.

We turn to the proof of the theorem. Suppose $f(x) =$

$g(x)h(x)$ over R, where $f(x)$ has integral coefficients. Then

$$c_f f^*(x) = c_0 g^*(x) c_h h^*(x),$$

where each of $f(x)$, $g(x)h(x)$ has been written in the form just discussed. So

$$f(x) = c_f f^*(x) = (c_0 c_h) g^*(x) h^*(x).$$

But, by Theorem 3.6, $g^*(x)h^*(x)$ is primitive. Moreover the decomposition of $f(x)$ in this form is unique, so $f^*(x) = g^*(x)h^*(x)$ and

$$f(x) = c_f g^*(x) h^*(x).$$

But $f(x)$ and $f^*(x)$ have integral coefficients, and $f^*(x)$ is primitive, so c_f must be a positive integer. This proves the theorem.

THEOREM 3.8. (*Eisenstein's irreducibility criterion*). *Let p be a prime and $f(x) = a_0 + a_1 x + \cdots + a_n x^n$ a polynomial with integral coefficients such that*

$$p \nmid a_n, \quad p^2 \nmid a_0; \quad p \mid a_i, \quad i = 0, 1, \ldots n-1.$$

Then $f(x)$ is irreducible over R.

If $f(x)$ factors over R, then by Theorem 3.7 it has factors with integral coefficients. Suppose that

$$f(x) = (b_m x^m + \cdots + b_0)(c_k x^k + \cdots + c_0),$$

where the b_i, c_j are integers and $m + k = n$, the degree of $f(x)$. Since $a_0 = b_0 c_0$ and $p^2 \nmid a_0$, not both b_0 and c_0 are divisible by p. But $p \mid a_0$, so $p \mid b_0$ or $p \mid c_0$. We may suppose that $p \mid c_0$, $p \nmid b_0$.

Now $a_n = b_m c_k$ is not divisible by p, so c_k is not divisible by it either. Consider the list of coefficients $c_0, c_1, \ldots, c_k$. There must be a smallest value of $r \leq k$ such that c_r is not divisible by p, but $c_0, c_1, \ldots, c_{r-1}$ are so divisible.

By the multiplication formula for polynomials

$$a_r = b_0 c_r + b_1 c_{r-1} + \cdots + b_r c_0 \,.$$

All the terms on the right except $b_0 c_r$ are divisible by p. So a_r is not divisible by it either. But by hypothesis only one of the coefficients a_i is not divisible by p, and that one is a_n. Then $r = n$. Since $r \leq k$, $n \leq k$. But $k + m = n$, so $n \geq k$. The two inequalities can be reconciled only if $n = k$.

Hence one of the proposed factors of $f(x)$ necessarily has the same degree as $f(x)$. Then $f(x)$ must be irreducible.

As an application of Eisenstein's criterion we shall prove the irreducibility over R of two important polynomials. First observe that a polynomial $f(x)$ is irreducible if and only if $f(x + 1)$ is irreducible. For $f(x + 1) = g(x)h(x)$ if and only if $f(x) = g(x - 1)h(x - 1)$.

Let p be a prime and consider the so-called *cyclotomic* polynomial

$$\frac{x^p - 1}{x - 1} = x^{p-1} + x^{p-2} + \cdots + 1.$$

This is irreducible over R if

$$\frac{(x + 1)^p - 1}{(x + 1) - 1} = \frac{(x + 1)^p - 1}{x}$$

is also. But the latter is of the form (why?)

$$x^{p-1} + p(x^{p-2} + \cdots) + p,$$

and the irreducibility follows directly from Theorem 3.8.

As another important example consider the polynomial

$$\frac{x^{p^2} - 1}{x^p - 1} = x^{p(p-1)} + x^{p(p-2)} + \cdots + x^p + 1.$$

Replacing x by $x + 1$ yields

$$x^{p(p-1)} + pq(x),$$

where $q(x)$ has integral coefficients and final term 1. Once again Eisenstein's criterion shows that the polynomial is irreducible over R.

THEOREM 3.9. *If p is a prime number then the polynomials*

$$x^{p-1} + x^{p-2} + \cdots + x + 1$$

and

$$x^{p(p-1)} + x^{p(p-2)} + \cdots + x^p + 1$$

are irreducible over R.

3. **Symmetric polynomials.** Let $x_1, \ldots, x_n$ denote independent variables. By a polynomial in $x_1, \ldots, x_n$ over F we mean a finite sum of the form

$$g(x_1, \ldots, x_n) = \sum_{i_1, \ldots, i_n} a_{i_1, \ldots, i_n} x_1^{i_1} \cdots x_n^{i_n},$$

where the a's are elements in F and the exponents are non-negative integers. For example, $6x_1 + x_2x_3 + \frac{1}{5}x_3^2x_2 + x_1x_3$ is a polynomial in x_1, x_2, x_3.

A polynomial $g(x_1, \ldots, x_n)$ is *symmetric* if it is unchanged by any of the $n!$ permutations of the variables $x_1, \ldots, x_n$. For example, when $n = 3$ the polynomials $x_1 + x_2 + x_3$ and $x_1x_2 + x_2x_3 + x_3x_1$ are symmetric.

Now let z be still another variable, and define

$$f(z) = (z - x_1)(z - x_2) \cdots (z - x_n)$$

$$= z^n - \sigma_1 z^{n-1} + \sigma_2 z^{n-2} - \cdots (-1)^n \sigma_n.$$

It is easily verified that

$$\sigma_1 = x_1 + x_2 + \cdots + x_n$$

$$\sigma_2 = x_1 x_2 + x_1 x_3 + \cdots + x_2 x_3 + \cdots + x_{n-1} x_n$$

$$\cdots \cdots$$

$$\sigma_i = \text{sum of all products of } i \text{ different } x_j$$

$$\cdots \cdots$$

$$\sigma_n = x_1 x_2 \cdots x_n \, .$$

The σ_i are called the *elementary* symmetric functions in $x_1, \ldots, x_n$.

It is easy to see that if $f(x_1, \ldots, x_n)$ and $g(x_1, \ldots, x_n)$ are symmetric polynomials, then so too are the sum $f(x_1, \ldots, x_n) + g(x_1, \ldots, x_n)$ and the product $f(x_1, \ldots, x_n)g(x_1, \ldots, x_n)$. We shall now show how any symmetric polynomial can be expressed as a polynomial in the elementary symmetric functions.

THEOREM 3.10. *Every symmetric polynomial in $x_1, \ldots, x_n$ over a field F can be written as a polynomial over F in the elementary symmetric functions $\sigma_1, \ldots, \sigma_n$. If the coefficients of the first polynomial are rational integers, the same is true of the second.*

For example, let $n = 3$. Then

$$x_1^2 + x_2^2 + x_3^2$$

$$= (x_1 + x_2 + x_3)^2 - 2(x_1 x_2 + x_2 x_3 + x_3 x_1)$$

$$= \sigma_1^2 - 2\sigma_2 \, .$$

It suffices to prove the theorem for a *homogeneous* symmetric polynomial P, i.e., one expressible as a sum of terms

$$h = a x_1^{k_1} x_2^{k_2} \cdots x_n^{k_n}$$

of constant total degree $k = k_1 + k_2 + \cdots + k_n$. We may assume that no two terms of P have the same set of exponents k_1, k_2, ..., k_n by combining like terms if necessary.

We define a lexicographic order among the terms of P, saying that h is *higher* than another term

$$h' = bx_1^{l_1}x_2^{l_2}\cdots x_n^{l_n}$$

if any of the following conditions hold: $k_1 > l_1$; or $k_1 = l_1$ but $k_2 > l_2$; or generally, $k_1 = l_1, \ldots, k_{i-1} = l_{i-1}$ but $k_i > l_i$ for some $i \leq n$. For example, the highest terms of the elementary symmetric functions $\sigma_1, \sigma_2, \ldots, \sigma_n$ are x_1, x_1x_2, ..., $x_1x_2 \cdots x_n$, respectively.

Suppose that h is the highest term of P. It is easy to see that $k_1 \geq k_2 \geq \cdots \geq k_n$. Let P' be another homogeneous symmetric polynomial in $x_1, x_2, \ldots, x_n$ and let h' be the highest term of P'. Then hh' is the highest term of PP'. This assertion is valid for any product of homogeneous symmetric polynomials. For example, the highest term of $\sigma_1^{a_1}\sigma_2^{a_2}\cdots\sigma_n^{a_n}$ is

$$x_1^{a_1+a_2+\cdots+a_n}x_2^{a_2+\cdots+a_n}\cdots x_n^{a_n}.$$

Now the highest term of

$$Q = a\sigma_1^{k_1-k_2}\sigma_2^{k_2-k_3}\cdots\sigma_{n-1}^{k_{n-1}-k_n}\sigma_n^{k_n}$$

is h. Thus $P_1 = P - Q$ is either identically zero or is a homogeneous symmetric polynomial of the same total degree k as P, but having highest term h_1, which is not as high as h.

Repeating the argument, we form a product Q_1 of the σ's whose highest term is h_1. Then $P_1 - Q_1$ is either zero or a homogeneous symmetric polynomial P_2 of total degree k and highest term h_2 not as high as h_1.

There are only a finite number of products of powers

$x_1, \ldots, x_n$ having total degree k. After a finite number of steps, the above process must end, and $P_m - Q_m = 0$ for some m. Thus

$$P = Q + Q_1 + \cdots + Q_m,$$

and P is a polynomial in $\sigma_1, \ldots, \sigma_n$.

The coefficients of the Q_i clearly lie in the same field F as the coefficients of the original polynomial P. Also, if P has rational integer coefficients, so does each of the Q_i.

Frequently we shall use the following corollary of Theorem 3.10 rather than the theorem itself.

THEOREM 3.11. *Let $f(x)$ be a polynomial of degree n over F with roots $r_1, r_2, \ldots, r_n$. Let $p(x_1, \ldots, x_n)$ be a symmetric polynomial over F. Then $p(r_1, \ldots, r_n)$ is an element of F.*

As an example, let $f(x) = 2x^2 - 7x + 7$, $F = R$ and $p(x_1, x_2) = x_1^2 + x_2^2 = \sigma_1^2 - 2\sigma_2$. The roots r_1, r_2 of $f(x)$ are $(7 \pm \sqrt{7}i)/4$, and we have

$$p(r_1, r_2) = (r_1 + r_2)^2 - 2r_1r_2 = \left(\frac{7}{2}\right)^2 - 2 \cdot \frac{7}{2} = \frac{21}{4},$$

which is a rational number, as predicted by the theorem.

To show that Theorem 3.11 follows from Theorem 3.10 is not difficult. By Theorem 3.10 $p(x_1, \ldots, x_n)$ is a polynomial over F in $\sigma_1, \sigma_2, \ldots, \sigma_n$. This means that $p(r_1, \ldots, r_n)$ is a polynomial in $r_1 + r_2 + \cdots, r_1r_2 + r_1r_3 + \cdots, r_1r_2 \cdots r_n$. But these expressions are simply the coefficients of $f(x)/a_n$ if we write

$$f(x) = a_n(x^n - b_{n-1}x^{n-1} + b_{n-2}x^{n-2} - \cdots \pm b_0),$$

and all the b_i are in F.

An important consequence of Theorem 3.11 is the following corollary.

COROLLARY 3.12. *Let $f(x)$ and $g(x)$ be polynomials over a field F, and let $\alpha_1, \ldots, \alpha_n$; $\beta_1, \ldots, \beta_k$ be their respective roots. Then the products*

$$h_1(x) = \prod_{j=1}^{k} \prod_{i=1}^{n} (x - \alpha_i - \beta_j)$$

$$h_2(x) = \prod_{j=1}^{k} \prod_{i=1}^{n} (x - \alpha_i\beta_j)$$

are polynomials in x with coefficients in F.

We can write

$$f(x) = a_n(x - \alpha_1)(x - \alpha_2) \cdots (x - \alpha_n),$$

where a_n is the leading coefficient of $f(x)$. Then

$$f(x - \beta_j)$$
$$= a_n(x - \alpha_1 - \beta_j)(x - \alpha_2 - \beta_j) \cdots (x - \alpha_n - \beta_j)$$
$$= a_n \prod_{i=1}^{n} (x - \alpha_i - \beta_j)$$

Hence

$$a_n^k h_1(x) = \prod_{j=1}^{k} f(x - \beta_j).$$

The product is a polynomial in x each of whose coefficients is symmetric in $\beta_1, \ldots, \beta_k$. So by Theorem 3.11 its coefficients are in F. If we divide both sides by a_n^k it follows that the coefficients of $h_1(x)$ are in F, since F is a field.

To prove the second part of the theorem, assume first that no β_j is zero. Then

$$f\left(\frac{x}{\beta_j}\right) = a_n\left(\frac{x}{\beta_j} - \alpha_1\right) \cdots \left(\frac{x}{\beta_j} - \alpha_n\right),$$

so

$$\beta_j^n f\left(\frac{x}{\beta_j}\right) = a_n(x - \alpha_1\beta_j)\cdots(x - \alpha_n\beta_j),$$

and therefore

$$a_n^k h_2(x) = \prod_{j=1}^{k} \beta_j^n f\left(\frac{x}{\beta_j}\right).$$

The remainder of the proof goes much as before.

Assume now that $\beta_k = \ldots = \beta_{k-s+1} = 0$. Setting $g(x) = x^s g_1(x)$, we can apply the preceding argument to show that

$$h_2(x) x^{-sn} = \prod_{j=1}^{k-s} \prod_{i=1}^{n} (x - \alpha_i\beta_j)$$

is a polynomial with coefficients in F. Then $h_2(x)$ is also in $F[x]$.

Problems

1. Show that the set of numbers $\{a + bi: a, b \text{ rational}\}$ forms a field. This field is denoted by $R(i)$.

2. Suppose $P(x)$ is an irreducible polynomial over a field F and $Q(x) \in F[x]$ and $Q(x) \mid P(x)$. What can you say about $Q(x)$?

3. Assume that $f(x)$ and $g(x) \in J[x]$ and $f(x) \mid g(x)$ in $J[x]$. For what $c \in J$ is $f(c) \mid g(c)$ true in J?

4. Find all $n \in J$ for which $3x^2 + 2nx + 12$ is irreducible over R.

5. Let $f(x)$ be a polynomial with real coefficients and assume $f(x) \geq 0$ for all real x. Prove that there exist polynomials $g(x)$ and $h(x)$ with real coefficients such that $f = g^2 + h^2$.
   Hint: The proof of Problem 2.8 is useful, as is the fact

that for any polynomial $P(x)$ with real coefficients $P(z) = 0$ if and only if $P(\bar{z}) = 0$.

6. Let $f(x)$ and $g(x) \in F[x]$. If $d(x) \in F[x]$ is a common divisor of $f(x)$ and $g(x)$ and if no other common divisor of $f(x)$ and $g(x)$ has greater degree than $d(x)$, then $d(x)$ is called a *greatest common divisor* of $f(x)$ and $g(x)$.

   (a) Show that $f(x)/d(x)$ and $g(x)/d(x)$ are relatively prime.

   (b) Show that there exist polynomials $s_0(x)$ and $t_0(x)$ in $F[x]$ such that

   $$d(x) = s_0(x)f(x) + t_0(x)g(x).$$

   (c) Show that $d(x)$ is unique to within units.

7. Show that if $f(x)$ and $g(x) \in F[x]$ and $f(x)$ and $g(x)$ are relatively prime over F, then $f(x)$ and $g(x)$ can have no roots in common.

8. Use the technique of the Euclidean algorithm (cf. Problem 1.2) to find the greatest common divisor over $R[x]$ of the polynomials

   $$P(x) = x^4 + 3x^3 + 10x^2 + 18x + 24$$

   $$Q(x) = x^4 + 2x^3 + 13x^2 + 12x + 42.$$

9. Find polynomials $P_1(x)$ and $P_2(x)$ in $J[x]$, each of degree two, such that all coefficients of the product $P_1(x)P_2(x)$ are even except the coefficient of x.

10. Let $P(x) = a_0 + \cdots + a_n x^n \in J[x]$. The *rational root theorem* asserts that any rational root of $P(x)$ has the form r/s, where $r \mid a_0$ and $s \mid a_n$. Show that this result is a special case of Theorem 3.7.

11. Show how the rational root theorem enables us to decide in a finite number of steps whether a given cubic polynomial in $J[x]$ is reducible. Use this cri-

terion to decide whether the following polynomials can be factored over R:

(a) $x^3 - x + 2$

(b) $x^3 - 12x^2 + 44x - 52$.

12. With the aid of Eisenstein's criterion or otherwise, test the following polynomials for irreducibility over R:

(a) $x^3 + 2x^2 + 8x + 2$ (d) $x^3 + 14$

(b) $x^3 + 2x^2 + 2x + 4$ (e) $5x^9 - 41$

(c) $x^3 + x^2 + x + 1$ (f) $x^2 + 5x + 25$

        (g) $5x^5 + 30x^4 + 42x^3 + 6x + 12$.

13. Let $f(x) \in J[x]$ and $f(x)$ be monic. Let p be a rational prime.

(a) Show that if $f(x)$ is irreducible over the field $J \pmod p$, then $f(x)$ is irreducible over R. (Acquaintance with finite fields—which are not number fields!—is assumed for this problem.)

(b) Show by example that the converse of the preceding assertion is false.

(c) Test the irreducibility over R of the following polynomials, using the criterion of part (a) and suitable small p:

     (i) $x^3 - 12x^2 + 44x - 52$

     (ii) $x^4 + x^3 + x^2 + x + 1$.

Hint for (ii): Find all irreducible quadratic polynomials over the field $J \pmod 2$.

14. Let $P(x) = x^4 - 2x^3 + 6x - 3$. Choose a suitable $b \in J$ and prime p such that Eisenstein's criterion can be applied to $P(x + b)$.

15. Show that $f(x) = x^4 + 1$ is irreducible over R. Show that $f(x)$ can be factored over $R(i)$.

16. Using induction, verify the values of $\sigma_1, \ldots, \sigma_n$, the elementary symmetric functions in n variables.

17. Write the most general symmetric polynomial in x_1, x_2, and x_3 having total degree 4.
    Hint: The proof of Theorem 3.10 gives a useful classification of symmetric polynomials.

18. Let $n \geq 3$. Express $x_1^3 + x_2^3 + \cdots + x_n^3$ in terms of the elementary symmetric functions σ_1, σ_2, and σ_3.

19. Suppose $f(x) = x^2 - 5x + 7$ has roots α and β. With the aid of symmetric functions find $\alpha + \beta$, $\alpha\beta$, $\alpha^2 + \beta^2$, $\alpha^3 + \beta^3$.

20. Find a polynomial having roots 1, -1, 2, and 3 by using elementary symmetric functions in the four roots.

21. (a) Find a fourth degree polynomial over J satisfied by $\sqrt{2} + i$.
    (b) Find a sixth degree polynomial over J satisfied by $\sqrt{2}\sqrt[3]{3}$.

22. Complete the proof of Corollary 3.12.

Chapter IV

ALGEBRAIC NUMBER FIELDS

1. Numbers algebraic over a field. Let F be a number field. A number θ is said to be *algebraic* over F if it satisfies a non-trivial polynomial equation

$$a_n x^n + a_{n-1} x^{n-1} + \cdots + a_0 = 0$$

with coefficients in F. θ need not belong to F. For example $\sqrt{2}$ satisfies $x^2 - 2 = 0$ over R, but $\sqrt{2}$ is not in R.

Suppose now that θ is algebraic over F, and consider all polynomials over F of which θ is a root. Let $p(x)$ be one of lowest degree. Since we can always divide out the leading coefficient, we may assume $p(x)$ to be monic. Then $p(x)$ is called a *minimal polynomial for θ over F*. $p(x)$ is clearly irreducible; otherwise θ would satisfy a polynomial of lower degree.

THEOREM 4.1. *If θ is algebraic over F, it has a unique minimal polynomial.*

Let $p(x)$ be a minimal polynomial, and $q(x)$ any other polynomial over F satisfied by θ. Then

$$q(x) = g(x)p(x) + h(x),$$

where $h(x) \equiv 0$ or $h(x)$ is of lower degree than $p(x)$. Let $x = \theta$. Since $p(\theta) = q(\theta) = 0$ we find $h(\theta) = 0$. Then $h(x) \equiv 0$; otherwise $p(x)$ would not be minimal. So $p(x) \mid q(x)$.

Now if $q(x)$ were any other minimal polynomial of θ over F, the same argument shows that $q(x) \mid p(x)$. Hence $p(x) = cq(x)$, and since both are monic, $p(x) = q(x)$, as asserted. Henceforth we shall refer to *the* minimal polynomial.

We have proved incidentally the

COROLLARY 4.2. *Any polynomial satisfied by θ over F contains the minimal polynomial of θ as a factor.*

COROLLARY 4.3. *If f(x) and g(x) are relatively prime over F they have no roots in common.*

For if θ were a common root, then by Corollary 4.2 the minimal polynomial of θ over F would divide both $f(x)$ and $g(x)$, contrary to the assumption that they have no common factor.

COROLLARY 4.4. *An irreducible polynomial of degree n over F has n distinct roots.*

For suppose the irreducible polynomial $f(x)$ has two roots which are the same. We can write

$$f(x) = a_n(x - r)^2 g(x).$$

Then, taking the derivative of each side,

$$f'(x) = a_n(x - r)^2 g'(x) + 2a_n(x - r)g(x),$$

so that $f'(r) = 0$. Now f is a constant multiple of the minimal polynomial of r by Corollary 4.2. But this cannot be, since $f'(x)$ is of lower degree than $f(x)$.

Let $θ$ be algebraic over F, and $p(x)$ its minimal polynomial, say of degree n. Then $θ$ is said to be of *degree n* over F. Let $θ_1, θ_2, \ldots, θ_n$ be the roots of $p(x)$, where $θ_1 = θ$. By Corollary 4.4 these n numbers are distinct. We call them the *conjugates of θ over F*. When $F = R$, we omit reference to the field and simply say *conjugates*.

An example. Let $F = R$. By Eisenstein's criterion $x^3 - 2$ is irreducible over R. Let $2^{1/3}$ denote the positive root. Then

$$2^{1/3}, \qquad 2^{1/3}\omega, \qquad 2^{1/3}\omega^2$$

are its conjugates, where $\omega = \frac{1}{2}(-1 + \sqrt{-3})$. For 1, ω, ω^2 are the roots of $x^3 - 1$.

THEOREM 4.5. *The totality of numbers algebraic over a field F forms a field.*

Let α and $\beta \neq 0$ be algebraic over F. We must show that $\alpha + \beta$, $\alpha - \beta$, $\alpha\beta$, α/β are themselves algebraic over F—that is, that they satisfy polynomials over F. Let $f(x)$ and $g(x)$ be the minimal polynomials over F for α and β respectively. Form the polynomials $h_1(x)$ and $h_2(x)$ described in Corollary 3.12. They are polynomials over F and are satisfied by $\alpha + \beta = \alpha_1 + \beta_1$ and $\alpha\beta = \alpha_1\beta_1$. Hence the sum $\alpha + \beta$ and the product $\alpha\beta$ are algebraic. Since $-\beta$ satisfies $g(-x)$, $-\beta$ is algebraic. Hence the sum $\alpha + (-\beta) = \alpha - \beta$ is algebraic. Finally, if m is the degree of $g(x)$, then $1/\beta$ satisfies $x^m g(1/x)$, so $1/\beta$ is algebraic. By the result for a product, $\alpha \cdot (1/\beta)$ is also algebraic over F.

Later we shall give an alternative proof of this theorem independent of symmetric functions.

2. Extensions of a field.

Let F be a field. Then any field K containing F is called an *extension* of F. Every number field, for example, is an extension of the field R of rational numbers.

If θ is algebraic over F, then $K = F(\theta)$ is defined to be the smallest field containing both F and θ. K is called a *simple algebraic extension* of F. Clearly K consists of all quotients $f(\theta)/g(\theta)$, where $f(x)$ and $g(x)$ are any polynomials over F for which $g(\theta) \neq 0$. In our next theorem we shall show that every element of $F(\theta)$ can be written more simply as a polynomial in θ.

THEOREM 4.6. *Every element α of $F(\theta)$ can be written uniquely in the form*

$$\alpha = a_0 + a_1\theta + \cdots + a_{n-1}\theta^{n-1} = r(\theta),$$

where the a_i are in F and n is the degree of θ over F.

Suppose, as we may, that $\alpha = f(\theta)/g(\theta)$, where $g(\theta) \neq 0$, and let $p(x)$ be the minimal polynomial for θ over F. Then $p(x)$ is irreducible and $p(x) \nmid g(x)$ (since otherwise $g(\theta) = 0$), so $p(x)$ and $g(x)$ are relatively prime. By Theorem 3.3 there exist polynomials $s(x)$ and $t(x)$ such that $s(x)p(x) + t(x)g(x) = 1$. Let $x = \theta$. Since $p(\theta) = 0$ we find that $1/g(\theta) = t(\theta)$, so that

$$\alpha = \frac{f(\theta)}{g(\theta)} = f(\theta)t(\theta)$$

is a polynomial in θ. For simplicity write $\alpha = h(\theta)$.

Now $h(x) = q(x)p(x) + r(x)$, where $r(x) \equiv 0$ or the degree of $r(x)$ is less than that of $p(x)$. Since $p(\theta) = 0$ it follows that

$$\alpha = h(\theta) = r(\theta).$$

Hence α is a polynomial in θ of degree at most $n - 1$.

It remains only to show that $r(x)$ is unique. Suppose also that $\alpha = r_1(\theta)$, where $r_1(x) \in F[x]$ and is of degree at most $n - 1$. Then $r(\theta) - r_1(\theta) = 0$ and θ satisfies the polynomial $r(x) - r_1(x)$. But θ satisfies no polynomial of degree less than n. It follows that $r_1(x)$ and $r(x)$ are identical.

Let $\alpha_1, \alpha_2, \ldots, \alpha_n$ be numbers algebraic over F. If $n > 1$, the smallest field $K = F(\alpha_1, \ldots, \alpha_n)$ containing F and the α_i is called a *multiple algebraic extension* of F.

THEOREM 4.7. *A multiple algebraic extension of F is a simple algebraic extension.*

To prove the theorem it is enough to prove that $F(\alpha, \beta)$ is simple when α and β are algebraic over F—that is, that $F(\alpha, \beta) = F(\theta)$ for some θ algebraic over F. For if $K = F(\alpha_1, \alpha_2, \alpha_3)$ we can write it $K = F(\alpha_1, \alpha_2)(\alpha_3)$ (why?) and apply the result twice; and similarly for other multiple extensions.

Let $\alpha_1, \ldots, \alpha_l$; $\beta_1, \ldots, \beta_m$ be the conjugates over F of α and β respectively; we number them so that $\alpha_1 = \alpha$ and $\beta_1 = \beta$. If $k \neq 1$ then $\beta_k \neq \beta$, since conjugates over F are distinct. Hence for each i and each $k \neq 1$ the equation

$$\alpha_i + x\beta_k = \alpha_1 + x\beta_1$$

has at most one solution for x in F. Since there are only a finite number of such equations and hence only a finite number of solutions x, we can choose a number $c \neq 0$ in F different from all solutions x. Then

$$\alpha_i + c\beta_k \neq \alpha + c\beta$$

for all i and all $k \neq 1$. Now let $\theta = \alpha + c\beta$. We shall show that $F(\theta) = F(\alpha, \beta)$ and this will prove the theorem.

First, every element in $F(\theta)$ lies in $F(\alpha, \beta)$, for each element in $F(\theta)$ can, according to Theorem 4.6, be written in the form

$$a_0 + a_1\theta + \cdots + a_{n-1}\theta^{n-1}$$
$$= a_0 + a_1(\alpha + c\beta) + \cdots + a_{n-1}(\alpha + c\beta)^{n-1},$$

and the right hand member is certainly in $F(\alpha, \beta)$.

We must show now that every element of $F(\alpha, \beta)$ lies in $F(\theta)$. This will be achieved if we can prove that α and β are in $F(\theta)$. For if they are, they are of the form $\alpha = r(\theta)$, $\beta = s(\theta)$, where $r(x)$ and $s(x) \in F[x]$. Every element of

$F(\alpha, \beta)$ is then of the form

$$\frac{u(\alpha, \beta)}{v(\alpha, \beta)} = \frac{u(r(\theta), s(\theta))}{v(r(\theta), s(\theta))}$$

where $u(x, y)$ and $v(x, y)$ are polynomials with coefficients in F. This quotient certainly is in $F(\theta)$. It is enough to show that β is in $F(\theta)$, for then $\alpha = \theta - c\beta$ is also. This we proceed to do now.

Let $f(x)$ and $g(x)$ be the minimal polynomials for α and β respectively. Since $f(\theta - c\beta) = f(\alpha) = 0$, the number β satisfies the equations $g(x) = 0$ and $f(\theta - cx) = 0$. $g(x)$ and $f(\theta - cx)$ have only the root β in common. For the roots of $g(x)$ are $\beta_1, \ldots, \beta_m$ and if $f(\theta - c\beta_i) = 0$ for some $i \neq 1$ then $\theta - c\beta_i$ would be one of the α_j, contrary to the choice of c.

Now $g(x)$ and $f(\theta - cx)$ are polynomials in x *with coefficients in* $F(\theta)$, and they have exactly one root β in common. Let $h(x)$ be the minimal polynomial for β over $F(\theta)$. By Corollary 4.2, $h(x) \mid g(x)$ and $h(x) \mid f(\theta - cx)$ in $F(\theta)[x]$. $h(x)$ cannot be of higher than the first degree, for otherwise $g(x)$ and $f(\theta - cx)$ would have more than one root in common. Hence $h(x) = \gamma x + \delta$, where γ and δ are in $F(\theta)$. But $h(\beta) = 0$, so $\beta = -\delta/\gamma$ is in $F(\theta)$, and we are done.

As an example, suppose it is required to write $R(\sqrt{3}, \sqrt[3]{2})$ as a simple extension $R(\theta)$. The conjugates of $\sqrt{3}$ are $\sqrt{3}$, $-\sqrt{3}$ and those of $\sqrt[3]{2}$ are $\sqrt[3]{2}$, $\sqrt[3]{2}\omega$, $\sqrt[3]{2}\omega^2$. In this case we can choose c to be 1, and $\theta = \sqrt{3} + \sqrt[3]{2}$. Then $R(\sqrt{3}, \sqrt[3]{2}) = R(\sqrt{3} + \sqrt[3]{2})$.

We shall now give two proofs of the very important fact that every element of a simple, and hence also of a multiple algebraic extension of F is algebraic over F.

For one of the proofs we shall use the theory of symmetric functions. For the other we shall use the following lemma from elementary algebra; a proof can be found in Paragraph 27 of Thomas' book listed in the bibliography.

LEMMA 4.8. *If $n < m$ and if the a_{ij} are in a field F, then the system of equations*

$$\sum_{j=1}^{m} a_{ij} x_j = 0, \qquad i = 1, 2, \ldots, n,$$

has a solution for $x_1, \ldots, x_m$ in F, where not all the x_j are zero.

THEOREM 4.9. *If θ is algebraic over F, so is every element α of $F(\theta)$. Also, the degree of α is not greater than the degree of θ.*

First proof. Let α belong to $F(\theta)$, where θ is of degree n over F. By Theorem 4.6 each of the powers α^i, $i = 0, 1, \ldots, n$, of α can be written

$$\alpha^i = \sum_{j=0}^{n-1} a_{ij} \theta^j,$$

where the a_{ij} are in F. By the preceding lemma we can find in F a set of numbers d_i, not all zero, such that

$$\sum_{i=0}^{n} a_{ij} d_i = 0, \qquad j = 0, 1, \ldots, n-1,$$

for the number of "unknowns" d_i is greater by one than the number of equations. Then

$$\sum_{i=0}^{n} d_i \alpha^i = \sum_{i=0}^{n} d_i \sum_{j=0}^{n-1} a_{ij} \theta^j$$

$$= \sum_{j=0}^{n-1} \theta^j \sum_{i=0}^{n} a_{ij} d_i = 0,$$

so that α satisfies the polynomial $d_n x^n + d_{n-1} x^{n-1} + \cdots + d_0$ over F.

Second proof. By Theorem 4.6, $\alpha = r(\theta)$. Let

$$f(x) = \prod_{i=1}^{n} (x - r(\theta_i)),$$

where $\theta_1, \theta_2, \ldots, \theta_n$ are the conjugates of θ over F. By Theorem 3.11 the coefficients of $f(x)$ as a polynomial in x are in F. Moreover $f(\alpha) = 0$, so that the proof is complete.

It is now possible to give a new proof of Theorem 4.5 independent of the theory of symmetric functions—as promised earlier. We must show that $\alpha + \beta$, $\alpha - \beta$, $\alpha\beta$, and α/β, $\beta \neq 0$, are algebraic over F when α and β are. Consider the field $F(\alpha, \beta)$, which contains these four elements in particular. It is a simple algebraic extension, by Theorem 4.7, and every element in it is algebraic over F, by Theorem 4.9.

3. **Algebraic and transcendental numbers.** A number θ is said to be an *algebraic number* if it is algebraic over the field R of rationals. According to Theorem 4.5 the totality of numbers algebraic over R forms a field. It is reasonable to ask whether this field coincides with the field of all complex numbers,—in other words, whether all numbers are algebraic numbers. We shall answer the question in the negative by exhibiting numbers which are not algebraic; such numbers are called *transcendental*.

LEMMA 4.10. *Let θ be a real algebraic number of degree $n > 1$ over R. There is a positive number M, depending on θ, such that*

$$\left| \theta - \frac{p}{q} \right| \geq \frac{M}{q^n}$$

for all rational numbers p/q with $q > 0$.

Let $f(x)$ be the primitive polynomial of lowest degree satisfied by θ; it differs at most by a multiplicative constant from the minimal polynomial for θ, and so is or degree n. Let M' be the maximum of $|f'(x)|$ in the interval $\theta - 1 \leq x \leq \theta + 1$, and let M be the smaller of 1 and $1/M'$. For this choice of M the desired inequality is valid. The proof has two parts.

First, suppose that $|\theta - p/q| \geq 1$. Then

$$\left| \theta - \frac{p}{q} \right| \geq M \geq \frac{M}{q^n}$$

for any rational integers p and $q \neq 0$, so we are done.

If $|\theta - p/q| < 1$, the proof is harder. By the law of the mean

$$\left| f(\theta) - f\left(\frac{p}{q}\right) \right| = \left| \theta - \frac{p}{q} \right| |f'(\xi)| \leq M' \left| \theta - \frac{p}{q} \right|,$$

where ξ lies between θ and p/q, and hence in the interval $(\theta - 1, \theta + 1)$. Moreover $f(\theta) = 0$, so

$$\left| f\left(\frac{p}{q}\right) \right| \leq M' \left| \theta - \frac{p}{q} \right|.$$

Now $f(p/q) \neq 0$; otherwise $f(x)$ would not be irreducible over R. Since $f(x)$ has integral coefficients and is of degree n, $|f(p/q)| = m/q^n$, where m is an integer. But $m \geq 1$, so that

$$\frac{1}{q^n} \leq \left| f\left(\frac{p}{q}\right) \right| \leq M' \left| \theta - \frac{p}{q} \right|.$$

Hence

$$\left| \theta - \frac{p}{q} \right| \geq \frac{1}{M'} \frac{1}{q^n} \geq \frac{M}{q^n},$$

by the choice of M.

We mention here a celebrated improvement of Lemma 4.10, due to K. F. Roth: *Let θ be a real algebraic number of degree $n > 1$ over R. If ρ is any real number greater than two, there exists a positive number δ depending on θ and ρ such that $|\theta - p/q| \geq \delta q^{-\rho}$ for all rational numbers p/q, $q > 0$.* A curious feature of the proof of Roth's Theorem is that it is "ineffective" in the sense that it gives no way of determining a value for δ. A proof of Roth's theorem is given in LeVeque, vol. II, Chapter 4. Subsequent research in this area has concerned itself with trying to make such estimations "effective."

THEOREM 4.11. (*Liouville*). *There exist transcendental numbers.*

Let $\xi = \sum\limits_{m=1}^{\infty} (-1)^m 2^{-m!}$,

and denote by

$$\xi_k = p_k/q_k = p_k/2^{k!}$$

the sum of the first k terms of the series for ξ. Then

$$\left|\xi - \frac{p_k}{q_k}\right| = 2^{-(k+1)!} - 2^{-(k+2)!} + \cdots$$

$$< 2^{-(k+1)!} < 2^{-k \cdot k!} = q_k^{-k}.$$

Suppose that ξ is algebraic of degree $n > 1$ over R. By the preceding inequalities

$$q_k^n \left|\xi - \frac{p_k}{q_k}\right| \leq q_k^{n-k}.$$

Let $k \to \infty$. Then

$$\lim_{k \to \infty} q_k^n \left|\xi - \frac{p_k}{q_k}\right| = 0.$$

From this we can obtain a contradiction. For, by the preceding lemma, there exists a number $M > 0$ such that

$$\left| \xi - \frac{p_k}{q_k} \right| \geq \frac{M}{q_k^n},$$

so $q_k^n \left| \xi - p_k/q_k \right| \geq M > 0$ for all k, contrary to the limit zero just obtained. Then ξ cannot be algebraic of degree $n > 1$.

It follows that ξ is either a rational number—that is, an algebraic number of degree 1—or is transcendental. We shall eliminate the first of these possibilities. Suppose $\xi = p/q$, where p and q are rational integers, $q > 0$. Choose an odd k so that $2^{k \cdot k!} > q$. Then the number defined by

$$\eta = 2^{k!}\xi q - 2^{k!}q \sum_{m=1}^{k} (-1)^m 2^{-m!} = 2^{k!}q \sum_{m=k+1}^{\infty} (-1)^m 2^{-m!}$$

is a positive rational integer. But

$$\eta < 2^{k!}q \frac{1}{2^{(k+1)!}} = \frac{q}{2^{k \cdot k!}} < 1,$$

by the choice of k. This contradiction leaves only the alternative that ξ is transcendental.

The number $\xi + \xi i$ is also transcendental. For if it were algebraic and therefore the root of a polynomial with real coefficients, its complex-conjugate $\xi - \xi i$ would also be a root. So the sum $(\xi + \xi i) + (\xi - \xi i) = 2\xi$ would be an algebraic number. This is impossible, since ξ is not an algebraic number.

For the reader familiar with the notion of denumerability a simpler proof of Theorem 4.11 is available. However, it does not yield any explicit examples of transcendental numbers. Briefly, it runs as follows. The totality of polynomials with rational coefficients is denumerable. Each has a finite number of roots, so the totality of algebraic numbers is denumerable. But the totality of complex numbers is non-denumerable, so that some of them must fail to be algebraic.

The problem of testing particular numbers for transcendence is a very difficult one. It was already known in the last century that e and π are transcendental. (Simple proofs can be found in Hardy and Wright or Landau, *Vorlesungen* III; see the bibliography.) But it is only since 1929 that numbers such as e^π and $2^{\sqrt{2}}$ have been shown to be transcendental. This is a consequence of a general theorem of Gelfond and Schneider which we state here without proof. An account of it is given by E. Hille in the American Mathematical Monthly vol. 49 (1942), pp. 654–661.

THEOREM 4.12. *Let α and β be algebraic numbers different from 0 and 1. If the number*

$$\eta = \frac{\log \alpha}{\log \beta}$$

is not rational, then it is transcendental.

We shall illustrate the theorem by proving from it that $2^{\sqrt{2}}$ is transcendental. Suppose, on the contrary, that $\alpha = 2^{\sqrt{2}}$ is algebraic. Let $\beta = 2$. Since

$$\eta = \frac{\log 2^{\sqrt{2}}}{\log 2} = \sqrt{2}$$

is irrational, then η must be transcendental. This is obviously false, so that α cannot be algebraic.

A similar argument proves that e^π is transcendental, provided we first observe that e^π can also be written i^{-2i}. Let the reader complete the proof.

In 1966 A. Baker found some far-reaching generaliza-
tions of the Gelfond-Schneider theorem. For example,
he proved

THEOREM 4.13. *Let* $\alpha_1, \ldots, \alpha_n$ *be algebraic numbers and
assume* $\log \alpha_1, \ldots, \log \alpha_n$ *are linearly independent over* R.
Then $1, \log \alpha_1, \ldots, \log \alpha_n$ *are linearly independent over
the algebraic numbers.*

If any of the α_i are non-positive, use the principal
branch of the logarithm. The notion of linear independence
is defined at the opening of the next chapter.

Surprisingly, by the very same method, Baker showed
how to obtain all integer solutions of certain equations,
for example, $y^2 = x^3 + k$, where k is any fixed rational
integer. References and further results in this area can be
found in a survey article by Serge Lang in the Bulletin of
the American Mathematical Society, vol. 77 (1971), pp.
635–677.

Problems

1. Give another proof of the uniqueness of the minimal
   polynomial by subtracting two (monic) minimal
   polynomials for a given θ.
2. Let θ be the root of a monic irreducible polynomial P
   over a field F. Show that P is the minimal polynomial
   for θ over F.
3. Find the minimal polynomial of $\sqrt[3]{2} + 1$ over R. Show
   that the polynomial you have found *is* minimal.
4. Let θ be algebraic over F. Prove that

$$I = \{g(x) \in F[x] : g(\theta) = 0\}$$

   is an *ideal* of $F[x]$, i.e. show that

   (a) if $g(x)$ and $h(x) \in I$, then $g(x) - h(x) \in I$,
   (b) if $g(x) \in I$ and $f(x) \in F[x]$, then $f(x)g(x) \in I$.

Also, show that I can be characterized as the collection of all multiples of the minimal polynomial of θ over F.

5. Assume that α is algebraic of degree $m > 1$ over a field F. Is it possible that $(x - \alpha)^n$ be a polynomial over F for some positive integer n?

6. Show that if $\alpha \neq 0$, then the conjugates of α are non-zero.

7. Find the minimal polynomial and the conjugates over R of the following algebraic numbers: $\sqrt{3}, 4, \sqrt{-3}/5, \sqrt[3]{5}$.

8. Let α and β be numbers algebraic over a field F. Let $\alpha \sim \beta$ mean that β is a conjugate of α over F. Prove that "$\sim$" is an *equivalence* relation, i.e.,

   (a) $\alpha \sim \alpha$
   (b) $\alpha \sim \beta$ implies $\beta \sim \alpha$, and
   (c) $\alpha \sim \beta, \beta \sim \gamma$ imply $\alpha \sim \gamma$.

9. Let θ_1 and θ_2 be conjugates over F. Show that for any polynomial $f(x)$ in $F[x]$, $f(\theta_1) = 0$ if and only if $f(\theta_2) = 0$.

10. Let θ be a non-real algebraic number. Prove that $\bar{\theta}$, the complex conjugate of θ, is one of the conjugates of θ over R.

11. Verify directly that $\sqrt{2} + \sqrt[3]{3}$ and $\sqrt{2}\sqrt[3]{3}$ are algebraic over R.

12. Verify Theorem 4.6 directly in the case where $\theta = \sqrt{2}$ and $F = R$.

13. Let ω denote a non-real cube root of 1. Theorem 4.6 asserts that there exist rational numbers a and b such that

$$(2 - \omega) / (\omega^2 + 3) = a + b\omega.$$

Find a and b by the method of undetermined coefficients.

14. Find the inverse of $1 + \sqrt[7]{3}$ in $R(\sqrt[7]{3})$.

15. Show with the aid of Theorem 4.6 that $\sqrt{2} \notin R(\sqrt{5})$.

16. Express $\sqrt[3]{2}$ and $\sqrt{2}$ in the form $a_0 + a_1\theta + \cdots + a_5\theta^5$, where $a_i \in R$ and $\theta = \sqrt{2} - \sqrt[3]{2}$.

17. Does $\sqrt{3}$ lie in $R(\omega, i)$? (ω a non-real cube root of 1.)

18. Let α and β be algebraic over F. Prove that $F(\alpha, \beta) = (F(\alpha))(\beta) = (F(\beta))(\alpha)$. (This fact is used in proving Theorem 4.7.)

19. Does i belong to $R(\sqrt{2} + i)$? A non-computational argument is preferred.

20. Show that $\sqrt{2} + \sqrt{3}$ is irrational using the fact that $R(\sqrt{2} + \sqrt{3}) = R(\sqrt{2}, \sqrt{3})$.

21. (a) Find an algebraic number θ such that $R(2^{1/3}, i) = R(\theta)$.

    (b) Express $(4 + i)2^{-1/3}$ as a polynomial in θ.

22. Find a non-zero solution in R of the system of equations

$$x_1 + 2x_2 + 3x_3 + 4x_4 = 0$$

$$5x_1 \qquad + x_3 + 8x_4 = 0$$

$$2x_1 + 3x_2 + 7x_3 \qquad = 0.$$

23. Suppose that α is algebraic of degree n over R. Show that

    (a) α^2 is algebraic of degree $\leq n$ over R,

    (b) $\alpha^{1/2}$ is algebraic of degree $\leq 2n$ over R.

    In each case, give examples of "$<$" and "$=$".

24. Assume that $\sin \alpha$ is algebraic over R. Show that $\cos \alpha$ and $\sin(\alpha/2)$ are also algebraic over R. Show that $\sin 15°$ is algebraic over R.

25. Suppose that $\theta^3 - \theta^2 - \theta - 1 = 0$. Find the minimal polynomial over R for $\theta^2 + 3$.

26. Prove that $|\sqrt{2} - a/b| \geq 1/(3b^2)$ for all positive rational integers a, b.

CHAPTER V

BASES

1. Bases and finite extensions. Let F be a number field and K an extension of it. A set of numbers $\alpha_1, \alpha_2, \ldots, \alpha_r$ in K is said to be *linearly dependent* (*over* F) if it is possible to find a set of numbers $c_1, c_2, \ldots, c_r$ in F, not all zero, such that

$$c_1\alpha_1 + c_2\alpha_2 + \cdots + c_r\alpha_r = 0.$$

Otherwise the numbers $\alpha_1, \alpha_2, \ldots, \alpha_r$ are called *linearly independent*.

A set of numbers $\beta_1, \beta_2, \ldots, \beta_s$ in K is said to form a *basis* for K over F if for each element β in K there exists a unique set of numbers $d_1, d_2, \ldots, d_s$ in F such that

$$\beta = d_1\beta_1 + d_2\beta_2 + \cdots + d_s\beta_s.$$

Observe that the β_i are linearly independent, for otherwise 0 has a representation

$$0 = e_1\beta_1 + e_2\beta_2 + \cdots + e_s\beta_s,$$

where not all the e_i are zero, and also the representation

$$0 = 0\cdot\beta_1 + 0\cdot\beta_2 + \cdots + 0\cdot\beta_s,$$

contrary to the requirement of uniqueness. Theorem 4.6 asserts that $1, \theta, \ldots, \theta^{n-1}$ is a basis for $F(\theta)$ over F, where n is the degree of θ over F.

LEMMA 5.1. *If K has a basis of s elements over F, then any t numbers in K, $t > s$, are linearly dependent over F.*

Let $\beta_1, \ldots, \beta_s$ be a basis for K and let $\alpha_1, \ldots, \alpha_t$ be t numbers in K. By the definition of a basis we can find numbers a_{ij} in F such that

$$\alpha_i = \sum_{j=1}^{s} a_{ij}\beta_j, \qquad i = 1, \ldots, t.$$

Since $t > s$ we can invoke Lemma 4.8 to conclude that there exist numbers c_i in F, not all zero, such that

$$\sum_{i=1}^{t} a_{ij}c_i = 0, \qquad j = 1, \ldots, s.$$

It follows that

$$\sum_{i=1}^{t} c_i\alpha_i = \sum_{i=1}^{t} c_i \sum_{j=1}^{s} a_{ij}\beta_j$$

$$= \sum_{j=1}^{s} \beta_j \sum_{i=1}^{t} a_{ij}c_i = 0,$$

so that the α_i are linearly dependent.

THEOREM 5.2. *If $\alpha_1, \alpha_2, \ldots, \alpha_t$ and $\beta_1, \beta_2, \ldots, \beta_s$ are both bases for K over F, then $s = t$.*

If $s \neq t$ we can suppose $t > s$. By the preceding lemma the α_i must be linearly dependent. This is impossible, since they form a basis.

We have shown that if K has a basis over F, every basis has the same number n of elements. n is called the *degree* of K over F, and K is called a *finite extension* of degree n over F. We write $n = (K/F)$.

LEMMA 5.3. *If K is a finite extension of degree n over F, then any n linearly independent elements in K form a basis.*

Let $\alpha_1, \ldots, \alpha_n$ be a set of n linearly independent elements of K. We shall show that every $\alpha \in K$ can be represented uniquely as

$$\alpha = d_1\alpha_1 + \cdots + d_n\alpha_n \qquad (d_1, \ldots, d_n \in F).$$

The uniqueness of such a representation follows directly from the definition of linear independence. By Lemma 5.1, the set $\alpha, \alpha_1, \ldots, \alpha_n$ is linearly dependent. Thus there exists a relation

$$c_0\alpha + c_1\alpha_1 + \cdots + c_n\alpha_n = 0 \qquad (c_0, \ldots, c_n \in F)$$

with not all $c_i = 0$. In particular, $c_0 \neq 0$ by the linear independence of $\alpha_1, \ldots, \alpha_n$. We solve the last equation for α and obtain the required representation with $d_i = -c_i/c_0 \in F$ for $1 \leq i \leq n$.

THEOREM 5.4. *If* $\alpha_1, \ldots, \alpha_n$ *is a basis for* K *over* F *and*

$$\beta_j = \sum_{i=1}^{n} a_{ij}\alpha_i, \qquad j = 1, 2, \ldots n,$$

where the a_{ij} *are in* F, *then* $\beta_1, \ldots, \beta_n$ *is also a basis if and only if the determinant* $|a_{ij}|$ *is not zero.*

First suppose $|a_{ij}| \neq 0$. By the preceding results it is enough to show that the β_j are linearly independent. Suppose $\sum_{j=1}^{n} c_j\beta_j = 0$, where the c_j are in F. Then

$$0 = \sum_{j=1}^{n} c_j \sum_{i=1}^{n} a_{ij}\alpha_i = \sum_{i=1}^{n} \alpha_i \sum_{j=1}^{n} c_j a_{ij}.$$

Since the α_i are linearly independent

$$\sum_{j=1}^{n} c_j a_{ij} = 0, \qquad i = 1, \ldots, n.$$

The determinant $|a_{ij}| \neq 0$, so that all the c_j must vanish.

Conversely, suppose $|a_{ij}| = 0$. Then the equations immediately preceding are known to have a solution with the c_j in F, and not all zero. Retracing our steps, we find

$$\sum_{j=1}^{n} c_j \beta_j = 0,$$

so that the β_j are not linearly independent.

2. **Properties of finite extensions.** Next we propose to show that finite extensions and simple algebraic extensions of a field are the same thing.

LEMMA 5.5. *If K is a finite extension of F then every element α of K is algebraic over F.*

Let $n = (K/F)$. By Lemma 5.1 the $n + 1$ numbers $1, \alpha, \alpha^2, \ldots, \alpha^n$ are linearly dependent, so that $c_0, c_1, \ldots, c_n$, not all zero, exist in F and satisfy

$$c_0 + c_1\alpha + \cdots + c_n\alpha^n = 0.$$

It follows that α satisfies a polynomial over F.

The preceding argument suggests a method for making an extension which is not finite. Let $F = R$ and let ξ be a transcendental number. Let K be the smallest field containing R and ξ. K is an extension of R, but not a finite one. Indeed, for any positive integer n, the set $1, \xi, \ldots, \xi^n$ is linearly independent over R.

THEOREM 5.6. *An extension K of F is finite if and only if it is a simple algebraic extension.*

First suppose that K is a finite extension of F, and let $\alpha_1, \ldots, \alpha_n$ be a basis. By the preceding lemma each α_i is algebraic over F. Then $K = F(\alpha_1, \ldots, \alpha_n)$. It follows from Theorem 4.7 that K is a simple algebraic extension of F.

Suppose conversely that $K = F(\theta)$, where θ is of degree n over F. By Theorem 4.6 the numbers $1, \theta, \ldots, \theta^{n-1}$

form a basis for K over F. This completes the proof. Note that (K/F) is the same as the degree of θ over F.

THEOREM 5.7. *If K is finite over F, and E over K, then E is finite over F. Moreover*

$$(E/F) = (E/K) \cdot (K/F).$$

Let $\alpha_1, \ldots, \alpha_n$ be a basis for K over F, and $\beta_1, \ldots, \beta_m$ for E over K. We shall show that the mn products $\alpha_i \beta_j$ form a basis for E over F.

If $\alpha \in E$, it can be expressed as $\alpha = \sum_{i=1}^{m} \gamma_i \beta_i$, where the γ_i are unique numbers in K, for the β_i form a basis of E over K. Similarly each γ_i can be written $\sum_{j=1}^{n} a_{ij} \alpha_j$, where the a_{ij} are unique numbers in F. Then we have the unique representation

$$\alpha = \sum_{i=1}^{m} \beta_i \sum_{j=1}^{n} a_{ij} \alpha_j = \sum_{i=1}^{m} \sum_{j=1}^{n} a_{ij} \alpha_j \beta_i ,$$

as required. The formula given in the theorem follows immediately.

We can now prove the following refinement of Lemma 5.5.

COROLLARY 5.8. *If K is of degree n over F, then any element α of K is algebraic over F, of degree dividing n.*

Let $E = F(\alpha)$. Then

$$n = (K/F) = (K/E)(E/F).$$

Hence (E/F) divides n. But, by the remark at the end of the proof of Theorem 5.6, the degree of E over F is the same as the degree of α over F.

THEOREM 5.9. *If α satisfies the equation*

$$\alpha_n x^n + \alpha_{n-1} x^{n-1} + \cdots + \alpha_0 = 0$$

where the α_i are algebraic over F, then α is algebraic over F.

Let $E = F(\alpha_0, \ldots, \alpha_n)$. This can be written as a simple algebraic extension of F. By Theorem 5.6, E is a finite extension of F. Moreover $E(\alpha)$ is a finite extension of E and hence, by Theorem 5.7, a finite extension of F. Then α lies in a finite extension of F. By the preceding corollary α is algebraic over F.

In much of what follows in this book the field F will be taken to be the field R of rational numbers. An *algebraic number field* is any *finite* (hence simple) *extension of R*. The totality of algebraic numbers, while it forms a field (Theorem 4.5), does not form an algebraic number field. For suppose this field were of degree n over R. The presence in it of an element of degree greater than n would contradict Corollary 5.8. But it is easy to produce an algebraic number of degree $n + 1$. The polynomial $x^{n+1} - 2$ is irreducible over R, by Eisenstein's criterion, so $2^{1/n+1}$ is of degree $n + 1$.

3. Conjugates and discriminants. The reader is reminded that the conjugates over F of a number α algebraic over F are the roots of the minimal polynomial of α over F. We find it useful to define a new concept of conjugacy, and to discuss its relation to the old.

Let $K = F(\theta)$ be a finite extension of degree n over F, and suppose α to be a number in K. By Corollary 5.8 the degree m of α over F divides n. According to Theorem 4.6 α can be written uniquely in the form

$$\alpha = \sum_{i=0}^{n-1} c_i \theta^i = r(\theta),$$

where the $c_i \in F$.

Let $\theta_1, \ldots, \theta_n$ be the conjugates of θ over F. Then the numbers

$$\alpha_i = r(\theta_i), \qquad i = 1, \ldots, n,$$

are called the *conjugates of α for $F(\theta)$*. So α has n conjugates in the new sense, but m in the old, where $m \mid n$.

It may appear at the moment that the numbers $\{\alpha_i\}_{i=1}^{n}$ depend on the choice of θ. However, we shall presently show that they depend only on the field K.

It is easily seen that the conjugates of $\alpha + \beta$ for $F(\theta)$ are $\alpha_1 + \beta_1, \ldots, \alpha_n + \beta_n$. Also, we now show that the conjugates of $\alpha\beta$ for $F(\theta)$ are $\alpha_1\beta_1, \ldots, \alpha_n\beta_n$. Let $\alpha = r(\theta)$, $\beta = s(\theta)$ where $r(x)$ and $s(x)$ are polynomials over F of degree $\leq n - 1$. Let $p(x)$ be the minimal polynomial of θ over F. By the division algorithm $r(x)s(x) = p(x)q(x) + t(x)$, where $t(x)$ has degree $\leq n - 1$. Now $p(\theta_i) = 0$ for $1 \leq i \leq n$, and we have $\alpha_i\beta_i = r(\theta_i)s(\theta_i) = t(\theta_i)$.

The relation between the old and the new conjugates is settled by our next theorem.

THEOREM 5.10. (i) *The conjugates of α for $F(\theta)$ are the conjugates over F each repeated n/m times.* (ii) *α is in F if and only if all conjugates for $F(\theta)$ are the same.* (iii) *$F(\alpha) = F(\theta)$ if and only if all its conjugates for $F(\theta)$ are distinct.*

As we showed in the second proof of Theorem 4.9, the polynomial

$$f(x) = \prod_{1}^{n} (x - r(\theta_i))$$

is a polynomial over F, and $f(\alpha) = 0$. ($f(x)$ is called the *field polynomial for α*). Let $g(x)$ be the minimal polynomial for α over F. By Corollary 4.2, $g(x) \mid f(x)$, so we can write

$$f(x) = [g(x)]^s h(x),$$

where $g(x)$ and $h(x)$ are relatively prime. We prove that $h(x) \equiv 1$. Note that if $h(x)$ is a constant at all it must be 1, since $g(x)$ and $f(x)$ are monic.

If $h(x)$ is not a constant, it has one of the $r(\theta_i)$, say $r(\theta_I)$, as a root. Then $h(r(x))$ vanishes when x equals θ_I. Let $p(x)$ be the minimal polynomial for θ, and hence for θ_I. Then $p(x) \mid h(r(x))$. It follows that $h(r(x))$ vanishes for all the θ_i, in particular for θ. So $h(r(\theta)) = h(\alpha) = 0$. This is impossible by Corollary 4.3, since $g(\alpha) = 0$ and $g(x), h(x)$ are relatively prime.

Hence $f(x) = [g(x)]^s$. Since m is the degree of α over F, $s = n/m$, the field polynomial is a power of the minimal polynomial. This proves (i).

As for (ii), if α is in F then $g(x) = x - \alpha$, $m = 1$, $s = n/m = n$, and $f(x) = (x - \alpha)^n$, so all the conjugates are the same. Conversely, if all the conjugates are the same $f(x) = (x - \alpha)^n$, so $s = n$, $m = 1$, and α is in F.

Finally, we prove (iii). Note that

$$\left(\frac{F(\theta)}{F}\right) = \left(\frac{F(\theta)}{F(\alpha)}\right) \cdot \left(\frac{F(\alpha)}{F}\right),$$

so $F(\theta) = F(\alpha)$ if and only if $m = n$, $s = 1$. In this case $f(x) = g(x)$ and all the conjugates are distinct. On the other hand, if the conjugates are distinct $s = 1$, $m = n$, and the result follows. The theorem is proved.

We now show that the conjugates of an element for $F(\theta)$ depend only on the field $F(\theta)$ and not on the particular choice of θ. Let $\theta^* \neq \theta$ and assume $F(\theta^*) = F(\theta)$. We can express $\theta = s(\theta^*)$ for some $s(x) \in F[x]$. Case (iii) of the preceding theorem asserts that θ has n distinct conjugates for $F(\theta^*)$. Thus there is a pairing of conjugates according to the rule $\theta_i = s(\theta_i^*)$, $1 \leq i \leq n$. For any $\alpha \in F(\theta)$ we have

$$\alpha = r(\theta) = r(s(\theta^*)) = t(\theta^*),$$

where $t(x)$ is the remainder resulting from dividing

$r(s(x))$ by the minimal polynomial of θ^*. Now $\alpha_i = r(\theta_i) = t(\theta_i^*)$, $1 \leq i \leq n$, and the α_i are conjugates of α for $F(\theta)$ and $F(\theta^*)$.

Now suppose $K = F(\theta)$ is of degree n over F, and let $\alpha_1, \ldots, \alpha_n$ be a basis. Denote by $\alpha_j^{(i)}$, $i = 1, \ldots, n$, the conjugates of α_j for K. The *discriminant* of the set $\alpha_1, \ldots, \alpha_n$ is defined by

$$\Delta[\alpha_1, \ldots \alpha_n] = |\alpha_j^{(i)}|^2,$$

where $|\alpha_j^{(i)}|$ is the determinant

$$\begin{vmatrix} \alpha_1^{(1)} & \alpha_2^{(1)} & \cdots & \alpha_n^{(1)} \\ \cdots & \cdots & \cdots & \cdots \\ \alpha_1^{(n)} & \alpha_2^{(n)} & \cdots & \alpha_n^{(n)} \end{vmatrix}.$$

The discriminant is well defined in that its value does not depend on the ordering of either the conjugates or the set $\alpha_1, \ldots, \alpha_n$.

If

$$\beta_k = \sum_{j=1}^{n} c_{jk}\alpha_j, \qquad k = 1, \ldots, n$$

is another basis, then $|c_{jk}| \neq 0$, by Theorem 5.4. We have

$$\beta_k^{(i)} = \sum_{j=1}^{n} c_{jk}\alpha_j^{(i)}, \qquad i, k = 1, \ldots, n,$$

and by the multiplication of determinants we arrive at the important formula

(5.1) $\quad \Delta[\beta_1, \ldots, \beta_n] = |c_{jk}|^2 \Delta[\alpha_1, \ldots, \alpha_n]$.

By Theorem 4.6 a particular basis for $F(\theta)$ is $1, \theta, \theta^2, \ldots, \theta^{n-1}$. If we use the fact that $(\theta^i)^{(j)}$ (the j-th conjugate of θ^i) is the same as $(\theta^{(j)})^i$ (the i-th power of $\theta^{(j)}$) we find that

$$D(\theta) = \Delta[1, \theta, \ldots, \theta^{n-1}]$$

$$= \begin{vmatrix} 1 & \theta^{(1)} & (\theta^{(1)})^2 & \cdots & (\theta^{(1)})^{n-1} \\ \cdots\cdots\cdots\cdots\cdots\cdots\cdots\cdots\cdots \\ 1 & \theta^{(n)} & (\theta^{(n)})^2 & \cdots & (\theta^{(n)})^{n-1} \end{vmatrix}^2 ,$$

and this Vandermonde determinant is known to have the value*

$$(5.2) \qquad D(\theta) = \prod_{1 \le i < j \le n} (\theta^{(i)} - \theta^{(j)})^2.$$

$D(\theta) \ne 0$ since the conjugates of θ for $F(\theta)$ are necessarily distinct. Since $D(\theta)$ is symmetric in the $\theta^{(i)}$ it is an element of F. It is obviously positive if all the $\theta^{(i)}$ are real. In (5.1) take $\alpha_i = \theta^{i-1}$, $i = 1, \ldots, n$. Then

$$\Delta[\beta_1, \ldots, \beta_n] = |c_{jk}|^2 D(\theta)$$

is an element of F. We have proved

THEOREM 5.11. *The discriminant of any basis for $F(\theta)$ is in F and is never zero. If F, θ and the conjugates of θ are all real then the discriminant of any basis is positive.*

4. **The cyclotomic field.** We shall now discuss a special kind of field which is of great importance. This will serve as an illustration of the theory which precedes, and will also be useful in our later work.

Let p be an odd prime. By Theorem 3.9 the cyclotomic polynomial $x^{p-1} + x^{p-2} + \cdots + 1$ is irreducible over R. Hence any root ζ generates a field $R(\zeta)$ of degree $p - 1$ over R. $R(\zeta)$ is called a *cyclotomic* field.

---

* Uspensky, Theory of Equations, p. 214. See also problem 24 below.

If ζ is any root then $\zeta, \zeta^2, \ldots, \zeta^{p-1}$ are all the roots because

(i) none of the ζ^s is 1, for otherwise ζ would satisfy a polynomial $x^s - 1$ of degree lower than $p - 1$;

(ii) they are all different, for the same reason; and

(iii) they all satisfy $x^p - 1 = 0$, since $(\zeta^s)^p - 1 = (\zeta^p)^s - 1 = 0$. This set of roots $\zeta, \ldots, \zeta^{p-1}$ are called the *primitive p^{th}* roots of unity. Since they lie on a circle of unit radius and none of them is ± 1, they are all non-real. The conjugates of ζ for $R(\zeta)$ are then simply $\zeta, \zeta^2, \ldots, \zeta^{p-1}$. Hence we can write $\zeta^{(i)} = \zeta^i$. We shall use this information to compute $D(\zeta)$.

By Theorem 4.6 with $n = p - 1$, a basis for $R(\zeta)$ is $1, \zeta, \ldots, \zeta^{p-2}$. By (5.2)

$$(5.3) \qquad D(\zeta) = \prod_{1 \leq i < j \leq p-1} (\zeta^i - \zeta^j)^2.$$

THEOREM 5.12. *If ζ is a primitive p^{th} root of unity, p an odd prime, then*

$$D(\zeta) = (-1)^{(p-1)/2} p^{p-2}.$$

Since $\zeta, \ldots, \zeta^{p-1}$ are all the primitive roots we have

$$(5.4) \qquad \frac{x^p - 1}{x - 1} = x^{p-1} + \cdots + 1 = \prod_{i=1}^{p-1} (x - \zeta^i).$$

Differentiate the right- and left-hand members, and let $x = \zeta^j$. Since $\zeta^p = 1$, we find that

$$(5.5) \qquad -\frac{p\zeta^{p-j}}{1 - \zeta^j} = \prod_{\substack{i=1 \\ i \neq j}}^{p-1} (\zeta^j - \zeta^i).$$

By (5.4) with $x = 0$ and $x = 1$ respectively it follows

that

$$\prod_{j=1}^{p-1} \zeta^{p-j} = \zeta \cdot \zeta^2 \cdots \zeta^{p-1} = (-1)^{p-1}$$

and

$$\prod_{j=1}^{p-1} (1 - \zeta^j) = p.$$

Hence by (5.5),

$$p^{p-2} = \prod_{j=1}^{p-1} \prod_{\substack{i=1 \\ i \neq j}}^{p-1} (\zeta^j - \zeta^i).$$

In the final product $i < j$ for half the factors and $j < i$ for the other half. There are $(p-1)(p-2)$ factors in all. Hence the last product is

$$p^{p-2} = (-1)^{(p-1)(p-2)/2} \prod_{1 \leq i < j \leq p-1} (\zeta^i - \zeta^j)^2.$$

But p is odd, so

$$(-1)^{(p-1)(p-2)/2} = (-1)^{(p-1)/2}.$$

If we combine these facts with the formula (5.3), the theorem follows.

Problems

1. Tell whether each of the following sets is linearly dependent or independent over R. Explain. As usual, ω denotes a non-real cube root of 1. (a) ω, $\sqrt{-3}$; (b) ω, ω^2, $\omega\sqrt{-3}$; (c) $2 + i$, $1 - 3i$, $12 + i$.
2. Suppose that K is an extension of a field F and α, β, $\gamma \in K$ and are linearly independent over F. Prove that the set $\alpha + \beta$, $\alpha - \gamma$, $\beta - \gamma$ is linearly independent over F.
3. Let K be an extension of F. Prove that $\beta_1, \ldots, \beta_n$,

elements of K, are linearly dependent over F if and only if one of the β's is expressible as a linear combination over F of the other β's.

4. Let $a, b, c, d \in J$. Show that a necessary and sufficient condition for $a + ib$, $c + id$ to be a basis for $R(i)$ over R is that $ad - bc = \neq 0$.

5. Let K be a finite extension of F and let α be a non-zero element of K. Show that there exists a basis for K over F containing α as an element.

6. Let K be a field containing F. Suppose that $\alpha_1, \ldots, \alpha_r$ are (not necessarily linearly independent) elements of K such that each α in K can be expressed as $\alpha = c_1\alpha_1 + \cdots + c_r\alpha_r$, where the $c_i \in F$. Prove that K is a finite extension of F. What is the relation between r and (K/F)?

7. Find a basis for $R(\sqrt{2})$ over R. Describe all possible bases for $R(\sqrt{2})$ over R in terms of the one you have found.

8. Show that $x^3 - 2$ is irreducible over R. Deduce that 1, $2^{1/3}$, $4^{1/3}$ is a linearly independent set over R.

9. Suppose that α and β are algebraic over K and $K(\alpha) = K(\beta)$. Show that α and β have the same degree over K.

10. Suppose that $\alpha \neq 0$ is algebraic over F, γ is transcendental over F, β is algebraic over $F(\alpha)$, δ is transcendental over $F(\alpha)$. Discuss the truth of the following assertions. Explain.

   (a) $\alpha + \beta$ is algebraic over F.
   (b) $1/\alpha$ is algebraic over F.
   (c) $\alpha + \delta$ is transcendental over F.
   (d) γ is transcendental over $F(\alpha)$.
   (e) $\gamma + \delta$ is transcendental over F.

11. Let α be the real positive number satisfying $\alpha^{13} = 2$. Is there a field F such that $R \subsetneqq F \subsetneqq R(\alpha)$?

12. Suppose that $p(x)$ is a cubic polynomial over R which is irreducible over R. Let K be an extension of R of degree 4. Prove that $p(x)$ is irreducible over K.

13. Prove that $R(i + \sqrt{2})$ has degree 4 over R and contains three subfields of degree 2 over R (cf. problem 4.19). Deduce that the polynomial found in problem 3.21(a) is minimal.

14. Let $K = R(\sqrt[3]{2},\ \sqrt{7},\ \sqrt{-5})$. Find the degree of K over R and exhibit a basis for K over R.

15. What is wrong with the following argument?

$$(R(\sqrt[3]{2}\omega)/R) = (R(\sqrt[3]{2})/R) = 3.$$

Therefore $(R(\sqrt[3]{2}\omega)/R(\sqrt[3]{2})) = 1$ and so $R(\sqrt[3]{2}\omega) = R(\sqrt[3]{2})$.

16. If $\cos\alpha$ is algebraic, prove that $\cos(\alpha/3)$ is algebraic. Hint. Show that $\cos(\alpha/3)$ is a root of the polynomial $4x^3 - 3x - \cos\alpha$.

17. It was shown in problem 4.19 that $i \in R(\sqrt{2} + i)$.

    (a) Express i as a polynomial in $\sqrt{2} + i$.
    (b) Compute the conjugates of i for $R(\sqrt{2} + i)$.

18. Prove that the conjugates of $2 - bi + 2\sqrt{2}i$ for $R(i + \sqrt{2})$ are distinct for any non-zero rational b.

19. Let $\theta = \sqrt[4]{5}$, $\beta = \sqrt{5}$, and $K = R(\theta)$.

    (a) Write the conjugates of θ.
    (b) Write the conjugates of β for K.
    (c) Find the field polynomial for β and express it as a power of the minimal polynomial of β over R.
    (d) Write the conjugates of $\gamma = 3 - 2\sqrt[4]{5}$ for K and determine the degree of $R(\gamma)$ over R.

20. Let $E = K(\kappa)$, $(E/K) = n$. Let $a \in K$, $a \neq 0$ and let $\lambda \in E$ and satisfy $\lambda\kappa = a$. Show that $\lambda_1, \ldots, \lambda_n$, the conjugates of λ for E, are distinct.

21. Let θ be algebraic and $\alpha \in F(\theta)$. Do the conjugates of α for $F(\theta)$ necessarily lie in $F(\theta)$?

22. Let D be a square-free rational integer. Find a basis γ, γ' for the field $R(\sqrt{D})$, where γ' is the conjugate of γ.

23. Let $\theta = \sqrt[4]{5}$ and $K = R(\theta)$.

   (a) Find $D(\theta) = \Delta[1, \theta, \theta^2, \theta^3]$.
   (b) Find $\Delta[\alpha_1, \alpha_2, \alpha_3, \alpha_4]$, where $\alpha_j = 1 + \theta^j$, $1 \leq j \leq 4$.
   (c) Find $\Delta[\beta_1, \beta_2, \beta_3, \beta_4]$ where $\beta_1 = 1$, $\beta_2 = \theta$, $\beta_3 = (1 + \theta^2)/2$, and $\beta_4 = (\theta + \theta^3)/2$.

24. Evaluate the Vandermonde determinant

$$\begin{vmatrix} 1 & x_1 & x_1^2 & \cdots & x_1^{n-1} \\ 1 & x_2 & x_2^2 & \cdots & x_2^{n-1} \\ \cdots\cdots\cdots\cdots\cdots\cdots\cdots\cdots \\ 1 & x_n & x_n^2 & \cdots & x_n^{n-1} \end{vmatrix}$$

using row and column operations and induction on n.

Chapter VI

ALGEBRAIC INTEGERS AND INTEGRAL BASES

1. Algebraic integers. Let $R(\theta)$ be an algebraic number field. What shall we mean by an *integer* in this field? With the example of the Gaussian integers as the "integers" in $R(i)$ before us, the following conditions seem reasonable to demand of our definition:

(i) The integers form a ring, i.e., if α and β are integers in $R(\theta)$, so are $\alpha + \beta$, $\alpha - \beta$ and $\alpha\beta$;

(ii) if α is an integer in $R(\theta)$ and is also a rational number, then it is a rational integer;

(iii) if α is an integer so are its conjugates; (in which of the two senses "conjugate" is to be taken is clearly a matter of indifference here.)

(iv) if $\gamma \in R(\theta)$, then $n\gamma$ is an algebraic integer for some non-zero rational integer n.

It turns out that the following definition meets all the requirements: an algebraic number is an *algebraic integer* if its minimal polynomial has only rational integers as coefficients. Since a minimal polynomial is monic α must satisfy an equation

$$p(x) = x^n + a_{n-1}x^{n-1} + a_{n-2}x^{n-2} + \cdots + a_0 = 0,$$

where the a_i are rational integers. It follows that the requirement (iii) is automatically fulfilled. To see that (ii) is also fulfilled is simple, for if α satisfies $p(x)$ and is rational, then its degree over R is 1, so $n = 1$, and so its minimal polynomial is simply $x + a_0 = 0$.

74

To prove that (i) holds is somewhat more complicated.

LEMMA 6.1. *If α satisfies any monic polynomial $f(x)$ with rational integral coefficients then α is an algebraic integer.*

Let $p(x)$ be the minimal polynomial for α over R. It is monic. We shall prove that *all* its coefficients are integers. It will follow that α is an algebraic integer.

By Corollary 4.2, $f(x) = p(x)q(x)$, where $q(x)$ is a polynomial over R. The proof of Theorem 3.7 shows that $f(x) = c_f p^*(x)q^*(x)$, where $p(x) = c_p p^*(x)$, and $p^*(x)$ and $q^*(x)$ are primitive. Since $f(x)$ is monic it is primitive, and $c_f = 1$. $p^*(x)$ and $q^*(x)$ have integral coefficients, and must therefore be monic, for their product $f(x)$ is monic. But $p(x)$ is also monic. Hence $c_p = 1$, and $p(x) = p^*(x)$ has integral coefficients.

THEOREM 6.2. *If $R(\theta)$ is an algebraic number field, then the integers in it form a ring.*

Let $\alpha_1, \ldots, \alpha_n ; \beta_1, \ldots, \beta_k$ be the conjugates over R of the algebraic integers $\alpha = \alpha_1$ and $\beta = \beta_1$ respectively. The elementary symmetric functions in $\beta_1, \ldots, \beta_k$ are rational integers since, except for sign, they are the coefficients of the minimal polynomial for the algebraic integer β. It follows from the second part of Theorem 3.10 that *any symmetric polynomial in $\beta_1, \ldots, \beta_k$ with rational integral coefficients is a rational integer.*

Now let $f(x)$ be the minimal polynomial for the integer α; and define

$$h(x) = \prod_{j=1}^{k} f(x - \beta_j).$$

This is a polynomial in x. Since $f(x)$ has integral coefficients, the coefficients of $h(x)$ are symmetric polynomials in the β_j with rational integral coefficients. By the italicized remark above, $h(x)$ has rational integers for coefficients. Since $f(x)$ is monic, so is $h(x)$. Finally

$$h(\alpha + \beta) = h(\alpha_1 + \beta_1)$$

$$= f(\alpha_1 + \beta_1 - \beta_1) \prod_{i=2}^{k} f(\alpha_1 + \beta_1 - \beta_j) = 0,$$

since $f(\alpha_1) = 0$. So by Lemma 6.1 $\alpha + \beta$ is an algebraic integer. It belongs to $R(\theta)$, since α and β do. The proofs for $\alpha - \beta$, $\alpha\beta$ are similar, and will be omitted (cf. the first proof of Theorem 4.5).

Note incidentally that this proof shows that $\alpha + \beta$, $\alpha - \beta$, $\alpha\beta$ are algebraic integers when α and β are, even if we do not suppose that α and β lie in the given field $R(\theta)$.

COROLLARY 6.3. *The totality of algebraic integers forms a ring.*

THEOREM 6.4. *If α satisfies an equation*

$$f(x) = x^n + \gamma_{n-1}x^{n-1} + \gamma_{n-2}x^{n-2} + \cdots + \gamma_0 = 0$$

where the γ_i are algebraic integers, then α is an algebraic integer.

Let $\gamma_j^{(i_j)}$ denote the conjugates of γ_j over R. Form the product

$$h(x) = \prod (x^n + \gamma_{n-1}^{(i_{n-1})}x^{n-1} + \cdots + \gamma_0^{(i_0)}),$$

over all these conjugates. When the product is multiplied out, the coefficient of a term x^k is

$$\sum_{j_1 + \cdots + j_\nu = k} \sum_{i_1} \cdots \sum_{i_\nu} \gamma_{j_1}^{(i_1)} \cdots \gamma_{j_\nu}^{(i_\nu)}$$

$$= \sum_{j_1 + \cdots + j_\nu = k} \prod_{\lambda=1}^{\nu} \left(\sum_{i_\lambda} \gamma_{j_\lambda}^{(i_\lambda)} \right).$$

The familiar argument on symmetric functions shows that

each of the inner sums on the right side of the above equation is in R, and thus the coefficient of x^k is in R. This coefficient is also an algebraic integer by property (i). We now apply property (ii) to conclude that all the coefficients of h are rational integers. Since $f(x) \mid h(x)$, $h(\alpha) = 0$. Finally $h(x)$ is monic, so Lemma 6.1 can be invoked to complete the proof.

We conclude this section by establishing property (iv).

THEOREM 6.5. *If θ is an algebraic number, there is a rational integer $r \neq 0$ such that $r\theta$ is an algebraic integer.*

θ satisfies an equation

$$a_n x^n + a_{n-1} x^{n-1} + \cdots + a_0 = 0,$$

where the a_i are rational integers. Then $a_n \theta$ satisfies

$$x^n + a_{n-1} x^{n-1} + a_n a_{n-2} x^{n-2}$$
$$+ a_n^2 a_{n-3} x^{n-3} + \cdots + a_n^{n-1} a_0 = 0.$$

This makes it an algebraic integer.

An elegant treatment of these elementary properties of algebraic integers without the use of symmetric functions will be found in the book of Landau listed in the bibliography.

2. **The integers in a quadratic field.** A *quadratic* field is a field of degree 2 over the rationals. Such a field is necessarily of the form $R(\theta)$, where θ is a root of a quadratic polynomial irreducible over the rationals. By Theorem 6.5 we can assume θ to be an algebraic integer; let it satisfy the equation $x^2 + bx + c = 0$, where b and c are rational integers. Then $\theta = (-b \pm \sqrt{b^2 - 4c})/2$. Remove from $b^2 - 4c$ all square factors, so that $(b^2 - 4c)/4 = s^2 D$ where s is rational and D is a rational integer containing no factor to higher than the first power.

Clearly $R(\theta) = R(\sqrt{D})$. In summary, every quadratic field is of the form $R(\sqrt{D})$, where D is a rational integer free of square factors.

By Theorem 4.6 the numbers 1, $\sqrt{D}$ form a basis for the field $R(\sqrt{D})$, so that every number in it can be written in the form $(l + m\sqrt{D})/n$, where l, m, n are rational integers. By cancelling, if necessary, we can assume that l, m and n are relatively prime, and that n is positive. We shall make this assumption.

How does one identify the algebraic integers among the elements of $R(\sqrt{D})$? The answer depends on the nature of the integer D. $(l + m\sqrt{D})/n$ is an integer if and only if it satisfies a quadratic equation $x^2 + bx + c = 0$, where b and c are rational integers, i.e.,

$$(6.1) \quad (l + m\sqrt{D})^2 + bn(l + m\sqrt{D}) + cn^2 = 0.$$

Equivalently,

$$(6.2) \qquad l^2 + m^2D + bnl + cn^2 = 0$$

and

$$m(2l + bn) = 0.$$

If $m = 0$ then $(l + m\sqrt{D})/n$ is an integer if and only if $n \mid l$; we assume then that $m \neq 0$. In this case $-2l = bn$, so that equation (6.2) becomes

$$m^2D - l^2 + cn^2 = 0.$$

Let $(l, n) = d$. Then $d^2 \mid m^2D$. Since D is square-free, $d \mid m$. But l, m and n by assumption share no factor except 1. Hence $d = 1$, and l and n are relatively prime. But $bn = -2l$, so that $n \mid 2$. Consequently $n = 1$ or 2.

If $n = 1$, then $(l + m\sqrt{D})/n$ is necessarily an integer. This is so since $l + m\sqrt{D}$ satisfies (6.1) if we take $b = -2l$ and $c = l^2 - m^2D$. The possibility $n = 2$ must

be scrutinized more closely. The number $(l + m\sqrt{D})/2$
satisfies the quadratic equation

$$x^2 - lx + \frac{l^2 - m^2 D}{4} = 0.$$

Consequently it is an integer if and only if $(l^2 - m^2 D)/4$
is a rational integer, that is

$$l^2 \equiv m^2 D (4).$$

Since $(l, n) = (l, 2) = 1$, l must be odd, say $l = 2t + 1$.
Then $l^2 = 4t^2 + 4t + 1$, and the requirement becomes

$$(6.3) \qquad\qquad 1 \equiv m^2 D (4).$$

This congruence holds if and only if m is odd and
$D \equiv 1 \pmod 4$. We have established

THEOREM 6.6. *Every quadratic field is of the form*
$R(\sqrt{D})$, *where D is a square-free rational integer. The
algebraic integers consist of these classes:*

1. *all numbers of the form $l + m\sqrt{D}$, where l and m are
rational integers, and*
2. *if $D \equiv 1 (4)$, but not otherwise, all numbers of the form
$(l + m\sqrt{D})/2$, where l and m are odd.*

3. Integral bases. Let $K = R(\theta)$ be an algebraic number field of degree n. By virtue of Theorem 6.5 we may
assume θ to be an integer and shall do so. By Theorem 4.6
every element of K can be written uniquely in the form
$\sum_{i=0}^{n-1} a_i \theta^i$, where the a_i are in R.

A set of integers $\alpha_1, \ldots, \alpha_s$ is called an *integral basis*
of K if every integer α in K can be written uniquely in
the form

$$\alpha = b_1 \alpha_1 + \cdots + b_s \alpha_s,$$

where the b_i are rational integers. We shall show that an integral basis is necessarily a basis.

Let β be an element of K. By Theorem 6.5 $r\beta$ is an integer for a suitable choice of the rational integer r. Consequently we can write

$$r\beta = b_1\alpha_1 + \cdots + b_s\alpha_s,$$

$$\beta = \frac{b_1}{r}\alpha_1 + \cdots + \frac{b_s}{r}\alpha_s.$$

It remains only to show that the α_i are linearly independent over R. Suppose

$$c_1\alpha_1 + \cdots + c_s\alpha_s = 0,$$

where the c_i are rational numbers. By multiplying the equation by the least positive common denominator we find a relation

$$d_1\alpha_1 + \cdots + d_s\alpha_s = 0,$$

where the d_i are rational integers. By the definition of an integral basis the d_i are all zero. Consequently the c_i are all zero, and the α_i are linearly independent.

LEMMA 6.7. *An integral basis is a basis.*

It follows immediately that $s = n$, that is, that the number of elements in an integral basis equals the degree of the field.

LEMMA 6.8. *If $\alpha_1, \ldots, \alpha_n$ is any basis of K over R consisting only of integers, then $\Delta[\alpha_1, \ldots, \alpha_n]$ is a rational integer.*

The conjugates of the α_i are algebraic integers. Consequently

$$\Delta = \Delta[\alpha_1, \ldots, \alpha_n] = \begin{vmatrix} \alpha_1^{(1)} & \cdots & \alpha_n^{(1)} \\ \cdots\cdots\cdots\cdots\cdots \\ \alpha_1^{(n)} & \cdots & \alpha_n^{(n)} \end{vmatrix}^2$$

is an algebraic integer. By Theorem 5.11 with $F = R$, Δ is also a rational number. So it is a rational integer.

THEOREM 6.9. *Every algebraic number field has at least one integral basis.*

Let $K = R(\theta)$ be an algebraic number field, where θ is assumed to be integral. Consider all bases for K whose elements are algebraic integers; $1, \theta, \ldots, \theta^{n-1}$ is an example*. Since, by Lemma 6.8, the discriminant of each such basis is a rational integer, there is some basis, $\omega_1, \ldots, \omega_n$, for which $|\Delta[\omega_1, \ldots, \omega_n]|$ is a minimum d. By Theorem 5.11, d is not zero.

We shall prove that $\omega_1, \ldots, \omega_n$ is an integral basis. For suppose it were not. Since it is in any case a basis, there is an integer ω, such that

$$\omega = a_1\omega_1 + \cdots + a_n\omega_n,$$

where the a_i are rational numbers, but not all integers. We may suppose that a_1 is not integral. Write it as $a_1 = a + r$, where a is a rational integer and $0 < r < 1$. Define

$$\omega_1^* = \omega - a\omega_1 = (a_1 - a)\omega_1 + a_2\omega_2 + \cdots + a_n\omega_n,$$

$$\omega_i^* = \omega_i, \qquad\qquad i = 2, \ldots, n.$$

The determinant

$$\begin{vmatrix} a_1 - a & a_2 & a_3 \cdots \\ 0 & 1 & 0 \cdots \\ 0 & 0 & 1 \cdots 0 \\ \cdots\cdots\cdots\cdots \\ 0 & \cdots & 1 \end{vmatrix} = a_1 - a = r$$

*We have not proved that $1, \theta, \ldots, \theta^{n-1}$ is an *integral* basis. That it need not be will become apparent later in the chapter.

is not zero. By Theorem 5.4 $\omega_1^*, \ldots, \omega_n^*$ is a basis; moreover it consists entirely of integers. Also

$$\Delta[\omega_1^*, \ldots, \omega_n^*] = r^2 \, \Delta[\omega_1, \ldots, \omega_n],$$

$$|\, \Delta[\omega_1^*, \ldots, \omega_n^*]\,| < |\, \Delta[\omega_1, \ldots, \omega_n]\,|$$

contrary to the choice of the last expression as a minimum. For reasons which are now clear an integral basis is also called a *minimal* basis.

THEOREM 6.10. *All integral bases for a field $K = R(\theta)$ have the same discriminant.*

Let $\alpha_1, \ldots, \alpha_n \, ; \beta_1, \ldots \beta_n$ be two integral bases. Then

$$\alpha_j = \sum_{i=1}^{n} c_{ij}\beta_i, \qquad j = 1, \ldots, n,$$

where the c_{ij} are rational integers. But

(6.4) $\qquad \Delta[\alpha_1, \ldots, \alpha_n] = |\, c_{ij}\,|^2 \, \Delta[\beta_1, \ldots, \beta_n],$

and $|\, c_{ij}\,|^2$ and the two discriminants are non-zero rational integers, so that

$$\Delta[\beta_1, \ldots, \beta_n] \mid \Delta[\alpha_1, \ldots, \alpha_n].$$

By reversing the roles of the α_j and β_i we find that

$$\Delta[\alpha_1, \ldots \alpha_n] \mid \Delta[\beta_1, \ldots, \beta_n].$$

Thus $\Delta[\alpha_1, \ldots, \alpha_n] = \pm\Delta[\beta_1, \ldots, \beta_n]$. By (6.4) the plus sign must prevail, and the proof is complete.

The discriminant d common to all integral bases is called the *discriminant of the field* K. Clearly $d \neq 0$. Since d is a rational integer, $|\, d\,| \geq 1$. It is known that if $K \neq R$, then $|\, d\,| > 1$.

4. Examples of integral bases. We begin by obtaining integral bases for the quadratic fields $R(\sqrt{D})$ discussed in §2.

First, if $D \not\equiv 1(4)$ then every integer is of the form $l + m\sqrt{D}$ (Theorem 6.6). Consequently an integral basis is 1, $\sqrt{D}$. Note that in this case the discriminant of the field is

$$d = \begin{vmatrix} 1 & \sqrt{D} \\ 1 & -\sqrt{D} \end{vmatrix}^2 = 4D.$$

Next suppose that $D \equiv 1(4)$. Every integer is of the form $(l + m\sqrt{D})/2$ where l and m are both even or both odd. In particular, $(1 + \sqrt{D})/2$ is an integer. It is easy to see that every integer can be expressed uniquely as

$$a + b\left(\frac{1 + \sqrt{D}}{2}\right),$$

where a and b are rational integers. (Cf. problem 8 below.) An integral basis is therefore 1, $(1 + \sqrt{D})/2$. Moreover

$$d = \begin{vmatrix} 1 & (1 + \sqrt{D})/2 \\ 1 & (1 - \sqrt{D})/2 \end{vmatrix}^2 = D.$$

THEOREM 6.11. *An integral basis for $R(\sqrt{D})$ is 1, $\sqrt{D}$ if $D \not\equiv 1(4)$ and 1, $(1 + \sqrt{D})/2$ if $D \equiv 1(4)$. In the former case $d = 4D$, in the latter $d = D$.*

A more complicated problem is the derivation of an integral basis for the cyclotomic field $R(\zeta)$, where ζ is a primitive p^{th} root of unity for p an odd prime. It was shown in §4 of Chapter V that the set 1, $\zeta, \ldots, \zeta^{p-2}$ is a basis for $R(\zeta)$. We shall now show that it is in fact an integral basis.

LEMMA 6.12. *If $\lambda = 1 - \zeta$, then 1, $\lambda, \ldots, \lambda^{p-2}$ is an integral basis for $R(\zeta)$.*

Let $\omega_1, \ldots, \omega_{p-1}$ be some integral basis for $R(\zeta)$.

Since ζ is an integer, so is each λ^j, and we can write

$$(6.5) \qquad \lambda^j = \sum_{i=1}^{p-1} c_{ij}\omega_i, \quad j = 0, \ldots, p-2,$$

where each c_{ij} is a rational integer. By (5.1)

$$(6.6) \quad \Delta[1, \lambda, \ldots, \lambda^{p-2}] = |c_{ij}|^2 \Delta[\omega_1, \ldots, \omega_{p-1}].$$

Now

$$1 = 1$$

$$(6.7) \qquad \lambda = 1 - \zeta$$

$$\lambda^2 = 1 - 2\zeta + \zeta^2$$

$$\lambda^3 = 1 - 3\zeta + 3\zeta^2 - \zeta^3$$

$$\cdots\cdots\cdots\cdots\cdots\cdots$$

so that

$$\Delta[1, \lambda, \ldots, \lambda^{p-2}] = |a_{ij}|^2 \Delta[1, \zeta, \ldots, \zeta^{p-2}].$$

Here a_{ij} is the matrix having zeros above the main diagonal and binomial coefficients elsewhere; in particular, the values on the diagonal are ± 1. Hence $|a_{ij}|^2 = +1$ and so by (6.6)

$$\Delta[1, \zeta, \ldots, \zeta^{p-2}] = |c_{ij}|^2 \Delta[\omega_1, \ldots, \omega_{p-1}].$$

Since $|c_{ij}|^2$ and $\Delta[\omega_1, \ldots, \omega_{p-1}]$ are rational integers, it follows from Theorem 5.12 that $|c_{ij}| = \pm p^k$ for some integer $k \geq 0$.

If we solve the system (6.5) for the ω_i, by Cramer's rule, they can be expressed in the form

$$(a_0 + a_1\lambda + \cdots + a_{p-2}\lambda^{p-2})/p^k,$$

where the a_i are rational integers. Since $\omega_1, \ldots, \omega_{p-1}$ is an integral basis, it follows that every integer in $R(\zeta)$ can be expressed in this form.

If $1, \lambda, \ldots, \lambda^{p-2}$ is *not* an integral basis there must therefore be an integer in $R(\zeta)$ of the form

$$\frac{a_0 + a_1\lambda + \cdots + a_{p-2}\lambda^{p-2}}{p}$$

where p does not divide all the numbers $a_0, a_1, \ldots, a_{p-2}$. Let a_m be the a_i with least subscript such that $p \nmid a_m$. Then

$$\frac{a_m\lambda^m + \cdots + a_{p-2}\lambda^{p-2}}{p}$$

is an algebraic integer, where $m \leq p-2$.

As we showed in §4 of Chapter V,

$$p = (1 - \zeta)(1 - \zeta^2) \cdots (1 - \zeta^{p-1})$$

$$= (1 - \zeta)(1 - \zeta) \cdots (1 - \zeta)\kappa$$

$$= \lambda^{p-1}\kappa = \lambda^{m+1}\kappa',$$

where κ and κ' are algebraic integers. Hence

$$\frac{a_m\lambda^m + \cdots + a_{p-2}\lambda^{p-2}}{\lambda^{m+1}}$$

is an algebraic integer. λ^{m+1} cancels into all terms but the first, so we can remove them to conclude that a_m/λ is an algebraic integer. We write $a_m = a$ for simplicity.

We shall prove that a/λ cannot be an algebraic integer, thus arriving at a contradiction. From this it will follow that $1, \lambda, \ldots, \lambda^{p-2}$ is an integral basis. Let $x = a/\lambda = a/(1 - \zeta)$. Then $\zeta = 1 - (a/x)$, so $1 = (1 - (a/x))^p$, $x^p = (x - a)^p$. Hence a/λ satisfies an equation

$$g(x) = px^{p-1} + p(\cdots) + a^{p-1} = 0,$$

where $p \nmid a$. Since $p \nmid a$, the polynomial $x^{p-1}g(1/x) = a^{p-1}x^{p-1} + p(\cdots) + p$ is irreducible by Eisenstein's criterion. Hence $g(x)$ is irreducible over R. Since it is primitive and its leading coefficient is not 1, its root a/λ is not an algebraic integer. This proves the lemma.

Now $1, \zeta, \ldots, \zeta^{p-2}$ is a basis consisting of algebraic integers and $D(\zeta) = D(\lambda)$. It follows from the fact that $1, \lambda, \ldots, \lambda^{p-2}$ is an integral basis and the proof of Theorem 6.9 that $1, \zeta, \ldots, \zeta^{p-2}$ is also an integral basis. Combining this result with Theorem 5.12 we arrive at

THEOREM 6.13. *The set* $1, \zeta, \ldots, \zeta^{p-2}$ *is an integral basis for* $R(\zeta)$. *This field has discriminant* $(-1)^{(p-1)/2}p^{p-2}$.

Problems

1. Show directly that the elements of G are algebraic integers. Conversely, show that if $\alpha \in R(i)$ and α is an algebraic integer, then $\alpha \in G$.

2. Let $f(x) = (x - \alpha_1\beta_1)(x - \alpha_1\beta_2)(x - \alpha_2\beta_1)(x - \alpha_2\beta_2)$. Verify that the coefficients of f can be written as sums of products of polynomials symmetric in α_1 and α_2 times ones symmetric in β_1 and β_2.

3. Find a polynomial over R which is satisfied by all the roots of $x^4 + \sqrt{3}x^2 + i$.

4. Show directly that a root of $x^3 + (2 - \sqrt{7})x + 1 + 3i$ is an algebraic integer.

5. Show how an algebraic integer can be expressed as the root of a primitive non-monic polynomial.

6. Let $\theta = (1/9)e^{2\pi i/7} + (1/5)\sqrt{2}$. Find a non-zero rational integer h for which $h\theta$ is an algebraic integer.

7. Factor $x^4 + 1$ over a suitable real quadratic field. (cf. problem 3.15)

8. Prove that

$$\{a + b\sqrt{D}: a, b \in J\}$$
$$\cup \{(2m + 1)/2 + (n + \tfrac{1}{2})\sqrt{D}: m, n \in J\}$$
$$= \{c + d(1 + \sqrt{D})/2: c, d \in J\}.$$

9. Give an example of a basis consisting of integers which is not an integral basis.

10. Let $\alpha_1, \alpha_2, \ldots, \alpha_n$ be algebraic integers which form a basis for a field $K \neq R$ and suppose that $\Delta[\alpha_1, \ldots, \alpha_n]$ is square free. Show that $\alpha_1, \ldots, \alpha_n$ is an integral basis for K.

11. For each of the fields $R(-5 + \sqrt{-28})$ and $R(1 - \sqrt{75/9})$ do the following:

   (a) Describe the integers.
   (b) Give an integral basis.
   (c) Describe all other integral bases.
   (d) Find the discriminant of the field.

12. Find the discriminant of $R(\alpha)$, where α is a root of $x^2 + 5x + 17$.

13. Does there exist an integral basis for the field $R(\sqrt{D})$ of the form γ, γ', where γ is an algebraic integer and γ' its conjugate when

   (a) $D = -3$,   (b) $D = -1$,   (c) $D = 5$?

   Exhibit such a basis when it exists.

14. Show that there exists no quadratic field with discriminant 19.

15. Let E and F be quadratic fields with the same discriminant. Show that $E = F$.

16. Let $a, b, c, d \in J$. Show that a necessary and sufficient condition for $a + ib, c + id$ to be an integral basis for $R(i)$ over R is that $ad - bc = \pm 1$.

CHAPTER VII

ARITHMETIC IN ALGEBRAIC NUMBER FIELDS

1. **Units and primes.** Consider the ring of *all* algebraic integers, and let us try to model a theory of factorization in this ring after the pattern of Chapter I. We might say that α *divides* β, written $\alpha \mid \beta$, if β/α is an algebraic integer. ϵ is a *unit* if ϵ divides 1. α is a *prime* if it is not zero or a unit, and if any factorization $\alpha = \beta\gamma$ into integers implies that either β or γ is a unit.

This attempt, natural in view of our earlier work, is unfortunately doomed to failure because there are no such primes in the ring of all algebraic integers. For let α be an integer different from zero or a unit. Then we can always write $\alpha = \sqrt{\alpha}\sqrt{\alpha}$. If α satisfies $p(x) = 0$, then $\sqrt{\alpha}$ satisfies $p(x^2) = 0$, so $\sqrt{\alpha}$ is an integer. This forces us to abandon the definitions just given.

Instead, let us confine our attention to the ring of all integers in a *fixed* algebraic number field $K = R(\theta)$. This is in fact what we did in Chapter I. The definitions given above will now have to be altered. α *divides* β, $\alpha \mid \beta$, if β/α is an integer of K. ϵ is a *unit* if $\epsilon \mid 1$. α is a *prime* if it is not zero or a unit, and if any factorization $\alpha = \beta\gamma$ into integers of K implies that either β or γ is a unit. Note that if ϵ and ϵ' are units, so are $1/\epsilon$ and $\epsilon\epsilon'$. Two integers α and β are *associates* or *associated* if α/β is a unit in K.

With these definitions factorization of integers in K into the product of primes is always possible. This we shall verify immediately. On the other hand, as we saw in

Chapter I, the ring H of all integers* in $R(\sqrt{-5})$ does not have the property of *unique* factorization. Before investigating the cause of this phenomenon and the method for remedying it, we shall prove that in K factorization into primes is possible, whether or not it is unique.

If α is an integer in K and K is of degree n over R, then α has n conjugates $\alpha_1, \ldots, \alpha_n$ for K. We define the *norm* of α, written $N(\alpha)$ or $N\alpha$, by

$$N\alpha = \alpha_1 \ldots \alpha_n.$$

Note that $N\alpha$ depends on the field K. For example $N2 = 2$ in R, but $N2 = 4$ in $R(i)$.

LEMMA 7.1. $N\alpha$ *is a rational integer.*

Let $f(x)$ be the field polynomial of α (as defined in the proof of Theorem 5.10). Since $f(x)$ is a power of the minimal polynomial it has integral coefficients. Hence

$$f(x) = x^n + a_{n-1}x^{n-1} + \cdots + a_0$$
$$= (x - \alpha_1)(x - \alpha_2) \cdots (x - \alpha_n),$$

where a_0 is a rational integer. Then

$$N\alpha = \alpha_1 \ldots \alpha_n = (-1)^n a_0$$

LEMMA 7.2. $N(\alpha\beta) = N\alpha \cdot N\beta$.

If $\alpha_1, \alpha_2, \ldots, \alpha_n; \beta_1, \beta_2, \ldots, \beta_n$ are the conjugates of α and β respectively for K, then $\alpha_1\beta_1, \alpha_2\beta_2, \ldots, \alpha_n\beta_n$ are the conjugates of $\alpha\beta$ for K. This implies the lemma.

LEMMA 7.3. α *is a unit in* K *if and only if* $N\alpha = \pm 1$.

For α is a unit if and only if $\alpha \mid 1$. If $\alpha \mid 1$ then $N\alpha \mid 1$,

---

* That H actually constitutes the ring of all integers in $R(\sqrt{-5})$ follows from Theorem 6.6, since $-5 \not\equiv 1(4)$.

$N\alpha = \pm 1$. If $N\alpha = \pm 1$, then $\alpha_1, \ldots, \alpha_n \mid 1$, and so $\alpha \mid 1$.

THEOREM 7.4. *If $N\alpha$ is a rational prime, α is prime in K.*
For if $\alpha = \beta\gamma$, $N\alpha = N\beta N\gamma$. Since $N\alpha$ is prime, one of $N\beta$ and $N\gamma$ is ± 1. Hence one of β and γ is a unit.

THEOREM 7.5. *Every integer in K, not zero or a unit, can be factored into a product of primes.*
If α is not already prime write $\alpha = \beta\gamma$, where neither β nor γ is a unit. Repeat the procedure for β and γ, and continue in this way. It must stop, for otherwise $\alpha = \gamma_1 \cdots \gamma_n$ where n is arbitrarily large, and then $\mid N\alpha \mid = \mid N\gamma_1 \mid \cdots \mid N\gamma_n \mid$ can be made as large as one pleases, since each factor $\mid N\gamma_i \mid$ exceeds unity.

COROLLARY 7.6. *There are an infinite number of primes in an algebraic field.*
The same argument used in §1 of Chapter II shows that there is an infinite number of primes in K if there is at least one. But there is at least one. For the number 2 certainly belongs to K, and by Theorem 7.5 it has a prime factor.
We shall resume the study of uniqueness of factorization in §3.

2. **Units in a quadratic field.** To illustrate some of the material of the preceding section we shall discuss the problem of determining the units in a quadratic field $R(\sqrt{D})$. If $\alpha = a + b\sqrt{D}$ is an integer in $R(\sqrt{D})$, then

$$N\alpha = (a + b\sqrt{D})(a - b\sqrt{D}) = a^2 - Db^2.$$

This reduces the problem of determining the units to the solution of the equation $a^2 - Db^2 = \pm 1$.
If $D \not\equiv 1(4)$ the integers are all the numbers of the form $l + m\sqrt{D}$, where l and m are rational integers. Then

to determine the units we must solve

(7.1) $$l^2 - Dm^2 = \pm 1$$

for rational integers l, m.

If $D \equiv 1(4)$ there are in addition to these the integers $(l + m\sqrt{D})/2$ where l and m are both odd. Then all further units come from the solution of

(7.2) $$l^2 - Dm^2 = \pm 4$$

in odd integers l, m.

Suppose first that $D < 0$. In this case the field $R(\sqrt{D})$ is called *imaginary*. A more accurate term is *non-real*, but the other is firmly entrenched. Note that the left-hand members of both (7.1) and (7.2) become positive, so the minus signs in the right-hand members must be dropped. Then the units are obtained from $l^2 - Dm^2 = 1$ and, if $D \equiv 1(4)$, also from $l^2 - Dm^2 = 4$, where l and m are odd. Since $-D > 0$, each of these equations can have at most a finite number of solutions. We shall determine them explicitly, first stopping to remind the reader that D is square-free.

If $D < -1$, then $l^2 - Dm^2 = 1$ has only the solutions $l = \pm 1, m = 0$; if $D < -4, l^2 - Dm^2 = 4$ has no solution with odd l and m. Hence if $D < -4$ the only units are ± 1. It remains to consider the cases $D = -1, -2, -3$. The first of these corresponds to the field $R(i)$, and we have already proved in Chapter I that the units in this field are $\pm 1, \pm i$. Secondly, since $D = -2 \not\equiv 1(4)$ the initial remark of this paragraph shows that the only units in $R(\sqrt{-2})$ are ± 1.

We turn our attention to $R(\sqrt{-3})$. Since $-3 \equiv 1(4)$ we can expect in addition to ± 1 further units arising from the solution in odd integers of $l^2 + 3m^2 = 4$. This has solutions $(1, 1)$, $(1, -1)$, $(-1, 1)$, $(-1, -1)$. So the units in $R(\sqrt{-3})$ are ± 1, $(1 \pm \sqrt{-3})/2$, $(-1 \pm \sqrt{-3})/2$. Note that they are the roots of $x \pm 1$ and $x^2 \pm x + 1$.

THEOREM 7.7. *The quadratic field $R(\sqrt{D})$ where D is negative and square-free, has only the units ± 1 unless $D = -1$, in which case there are the additional units $\pm i$, or unless $D = -3$ in which case there are the additional units*

$$\frac{1 \pm \sqrt{-3}}{2}, \qquad \frac{-1 \pm \sqrt{-3}}{2}.$$

But what if $D > 0$, so that the field is real? The situation becomes more complicated than in the imaginary case, and we shall content ourselves for the present with a solution for the case $D = 2$. Since $2 \not\equiv 1(4)$ only the solutions of (7.1)—that is, $l^2 - 2m^2 = \pm 1$—concern us.

LEMMA 7.8. $R(\sqrt{2})$ *has no unit between* 1 *and* $1 + \sqrt{2}$.

For suppose that $\epsilon = x + y\sqrt{2}$, where $x^2 - 2y^2 = \pm 1$, lies between 1 and $1 + \sqrt{2}$. Then $1 < \epsilon < 1 + \sqrt{2}$ and since $x - y\sqrt{2} = \pm 1/(x + y\sqrt{2})$, $-1 < x - y\sqrt{2} < 1$. Adding these inequalities we get $0 < 2x < 2 + \sqrt{2}$, $0 < x < 1.8$. Since x is an integer, $x = 1$. But then $1 < 1 + y\sqrt{2} < 1 + \sqrt{2}$, which is not possible for any integer y.

Observe that one solution of $l^2 - 2m^2 = \pm 1$ is $(1, 1)$, so that $\lambda = 1 + \sqrt{2}$ is a unit.

THEOREM 7.9. $R(\sqrt{2})$ *has an infinite number of units. They are given by* $\pm \lambda^n$, $n = 0, \pm 1, \pm 2, \ldots$.

To prove this note first that all the elements of $R(\sqrt{2})$ are real. Hence if ϵ is a unit in $R(\sqrt{2})$ it is positive or negative.

Suppose $\epsilon > 0$. Since $\lambda = 1 + \sqrt{2}$ exceeds 1 we can find an integer n such that $\lambda^n \leq \epsilon < \lambda^{n+1}$. Now $\epsilon\lambda^{-n}$ is a unit satisfying $1 \leq \epsilon\lambda^{-n} < 1 + \sqrt{2}$. By Lemma 7.8, $\epsilon = \lambda^n$. Now suppose $\epsilon < 0$. Since $-\epsilon$ is a positive unit, the proof is complete.

3. **The uniqueness of factorization***. It has already been observed that Theorem 7.5 says nothing about uniqueness (that is, to within order and units) of factorization into primes. In order to understand how the failure of uniqueness can come about, let us examine closely the integers in $R(\sqrt{-5})$. As we have seen in Chapter I,

$$21 = 3\cdot 7 = (1 + 2\sqrt{-5})(1 - 2\sqrt{-5}),$$

where all the factors which appear are primes. We then have the following situation: the number 3, a prime, divides $(1 + 2\sqrt{-5})(1 - 2\sqrt{-5})$, but fails to divide either factor in $R(\sqrt{-5})$. That this circumstance cannot come to pass in R or $R(i)$ was already proved in Chapter I.

In order to explain this situation we restore temporarily the definition of division given at the beginning of §1, but which was subsequently abandoned. Let $\alpha = 1 + 2\sqrt{-5}$, $\lambda = 2 + \sqrt{-5}$. Then

$$\frac{\alpha^2}{\lambda} = -2 + 3\sqrt{-5}, \qquad \frac{9}{\lambda} = 2 - \sqrt{-5}$$

are integers of $R(\sqrt{-5})$. It follows that their square roots $\alpha/\sqrt{\lambda}$, $3/\sqrt{\lambda}$ are integers, but these integers are not in $R(\sqrt{-5})$ (why?). In other words 3 and $1 + 2\sqrt{-5}$ are both divisible (in the extended sense of "division") by an

---

* The material of this section is adapted from Chapter V of Hecke's book listed in the bibliography.

integer $\sqrt{\lambda}$ which is not in $R(\sqrt{-5})$. Moreover, since

$$\sqrt{\lambda} = \left(-\frac{2\alpha}{\sqrt{\lambda}}\right)\alpha - \left(\frac{12 - 3\sqrt{-5}}{\sqrt{\lambda}}\right)3,$$

any other factor common to 3 and $\alpha = 1 + 2\sqrt{-5}$ divides $\sqrt{\lambda}$.

Similarly 7 and $1 - 2\sqrt{-5}$ have the "highest common factor" $\sqrt{\kappa}$, where $\kappa = 2 + 3\sqrt{-5}$.

A simple computation shows that

$$1 + 2\sqrt{-5} = \sqrt{\lambda}\sqrt{-\bar{\kappa}} \qquad 3 = \sqrt{\lambda}\sqrt{\bar{\lambda}}$$
$$1 - 2\sqrt{-5} = \sqrt{\bar{\lambda}}\sqrt{-\kappa} \qquad 7 = \sqrt{\kappa}\sqrt{\bar{\kappa}},$$

where the bar denotes the complex-conjugate. Then 21 can be factored, but not in $R(\sqrt{-5})$, as

$$21 = \sqrt{\lambda}\ \sqrt{\bar{\lambda}}\ \sqrt{-\kappa}\ \sqrt{-\bar{\kappa}},$$

and the various factorizations obtained in $R(\sqrt{-5})$ come from pairing these four factors in different ways.

In summary:

1. Prime numbers in $R(\sqrt{-5})$ which are not associated (that is, whose ratio is not a unit) can have a common factor which is not in $R(\sqrt{-5})$.

2. The totality of integers in $R(\sqrt{-5})$ which are divisible by a prime number α in $R(\sqrt{-5})$ need not coincide with the totality of integers in $R(\sqrt{-5})$ which are divisible by a factor of α not in $R(\sqrt{-5})$ and not a unit. ($\alpha = 1 + 2\sqrt{-5}$ is prime and $\sqrt{\lambda}$ divides both α and 3 but 3 is not divisible by α).

It appears then that in an algebraic number field K the primes are not necessarily the atoms from which all the integers are constructed. In $R(\sqrt{-5})$, for example it seems to be necessary to enlarge the ring of integers to include such "ideal" numbers as $\sqrt{\lambda}$, $\sqrt{\bar{\kappa}}$ which do not originally belong to it. But how shall we characterize those numbers which must be added to K?

Suppose an integer ξ is a possible candidate for admission to K by virtue of being a common factor to two integers relatively prime in K. Consider the *totality A of all integers in K which are divisible by ξ* (in the extended sense). It has the following property: if α and β are integers in A, so are all integers in K of the form $\lambda\alpha + \mu\beta$, where λ and μ are also integers in K. Any set of integers in K with the latter property we call an *ideal*.

This suggests the following procedure for answering the question raised above. Let us consider any ideal in K and try to prove that it is identical with the totality of integers in K which are divisible by some fixed integer ξ not necessarily in K. If we can accomplish this and can show further that ξ is in some sense unique, then we have characterized the missing integers by means of the ideals. This is the attack we shall pursue in the succeeding chapter.

But this poses another problem. If we are going to make ideals a substitute for integers, then the problem of factorization of integers is shifted to that of factorization of ideals. As we shall see, there is a completely satisfactory arithmetic for ideals, and by means of it we shall finally settle the problem of unique factorization.

4. Ideals in an algebraic number field. Let K be an algebraic number field. A set A of integers in K is an *ideal* in K if, together with any pair of integers α and β in A, the set also contains $\lambda\alpha + \mu\beta$ for any integers λ and μ in K. A set of integers $\omega_1, \ldots, \omega_r$ in A is said to form a *basis* for A if every element α of A can be uniquely represented in the form

$$(7.3) \qquad \alpha = c_1\omega_1 + \cdots + c_r\omega_r,$$

where the c_i are rational integers.

Let us denote by (0) the ideal consisting of 0 alone. We shall show that if an ideal $A \neq (0)$ in a field K has a basis $\omega_1, \ldots, \omega_r$, then r must equal n, the degree of the field. By virtue of the uniqueness of the representation (7.3) the set $\omega_1, \ldots, \omega_r$ must be linearly independent over R. Hence $r \leq n$, by Lemma 5.1. To show that $r < n$ is impossible, let $\beta_1, \ldots, \beta_n$ be an integral basis for K over R. If α is an element of A different from zero, then $\alpha\beta_1, \ldots, \alpha\beta_n$ are linearly independent and belong to A. On account of their linear independence they form a basis for K. Moreover

$$\alpha\beta_j = \sum_{i=1}^{r} a_{ij}\omega_i, \qquad j = 1, \ldots, n,$$

where the a_{ij} are rational integers. If $n > r$, then as in the proof of Lemma 5.1, there exist rational numbers $c_1, \ldots, c_n$, not all zero, such that

$$\sum_{j=1}^{n} a_{ij}c_j = 0 \qquad i = 1, \ldots, r.$$

Then $\sum_{j=1}^{n} \alpha c_j\beta_j = 0$, which is impossible since $\beta_1, \ldots, \beta_n$ form a basis for K over R.

To prove that an ideal $A \neq (0)$ necessarily has a basis we can imitate the proof of Theorem 6.9. Consider all sets $\alpha_1, \ldots, \alpha_n$ of integers in A which form a basis for K; the numbers $\alpha\beta_1, \ldots, \alpha\beta_n$ which occur above furnish an example. By Lemma 6.8, $\Delta[\alpha_1, \ldots, \alpha_n]$ is always a rational integer not zero, so we can pick such a set $\omega_1, \ldots, \omega_n$ from A for which $|\Delta[\omega_1, \ldots, \omega_n]|$ is a minimum. This is a basis for the ideal A, by precisely the same argument as that used to prove Theorem 6.9.

Conversely, every integer in K of the form (7.3) is in A. This is a consequence of the definition of an ideal and the fact that all rational integers are integers in K. Thus we have proved

THEOREM 7.10. *If K is of degree n over R and $A \neq (0)$ is an ideal in K, then there exist integers $\omega_1, \ldots, \omega_n$ in A such that A is the totality of integers of the form $\sum_{i=1}^{n} c_i \omega_i$, the c_i being unique rational integers.*

An ideal A is said to be *generated* by $\alpha_1, \ldots, \alpha_t$, written $A = (\alpha_1, \ldots, \alpha_t)$, if A consists of all sums $\sum_{i=1}^{t} \lambda_i \alpha_i$, where the λ_i are integers, *not* necessarily rational, in K. Obviously if $\omega_1, \ldots, \omega_n$ is a basis for A, then $A = (\omega_1, \ldots, \omega_n)$; but if $A = (\alpha_1, \ldots, \alpha_t)$, the α_i do not necessarily form a basis for A. For example, consider the ideal (2) in $R(i)$. This ideal consists of all integers of the form $2a + 2bi$, where a and b are rational integers; so a basis for (2) is 2, $2i$. The number 2 alone is not a basis for (2).

An ideal A is *principal* if it is generated by a single integer—that is, $A = (\alpha)$.

THEOREM 7.11. *Every ideal in R or in $R(i)$ is principal. There is an ideal in $R(\sqrt{-5})$ which is not principal.*

First, let A be an ideal in R. A consists entirely of rational integers. Suppose $A \neq (0)$; then it contains an element $a \neq 0$. In addition it contains $-a = (-1)a$. So both $\pm a$ belong to A, and one of these must be positive. Hence A contains positive integers. Let m be the least positive integer in A. If n is any other number in A we can find q and r such that

$$n = mq + r, \qquad 0 \leq r < m.$$

But every number $ns + mt$ is in A, and $r = n - mq$ in particular. Then $0 < r < m$ is impossible, by the choice of m as the least positive integer in A. Hence $r = 0$ and $n = mq$. In other words, every element of A is a multiple of m. Moreover, every multiple of m is in A, so $A = (m)$, as required.

A similar argument applies to the ideals A in $R(i)$; but instead of choosing the least positive number in A, we take an element of least positive norm, and apply Theorem 1.6.

On the other hand the ideal $B = (3, 1 + 2\sqrt{-5})$ is not principal in $R(\sqrt{-5})$. For if $B = (\beta)$, then $\beta \mid 3$, $\beta \mid (1 + 2\sqrt{-5})$. Since 3 and $1 + 2\sqrt{-5}$ are both prime in $R(\sqrt{-5})$ and are not associates, β must be a unit. The only units in $R(\sqrt{-5})$ are ± 1, so $B = (1)$. Thus the ideal B consists of *all* integers in $R(\sqrt{-5})$. However, the integer $\sqrt{-5}$ is not in B, since otherwise there would exist rational integers a, b, c, d so that

$$\sqrt{-5} = (a + b\sqrt{-5})3 + (c + d\sqrt{-5})(1 + 2\sqrt{-5}).$$

Multiplying out and equating real and imaginary parts, we obtain the simultaneous equations

$$3a + c - 10d = 0, \qquad 3b + 2c + d = 1.$$

These equations have no solutions in integers since 3 divides the sum of the left hand sides but $3 \nmid 1$.

The reader may suspect from the last theorem that unique factorization of integers in an algebraic number field is equivalent to the principality of all ideals in it. This conjecture will be confirmed in Chapter IX.

Problems

1. Let $a, b \in J$ and let K be a field containing R. Show that $a \mid b$ in K if and only if $a \mid b$ in R.

2. Let α and β be integers of a field $K = F(\theta)$ and let $E = F(\theta_1, \ldots, \theta_n)$, where $\theta_1, \ldots, \theta_n$ are the conjugates of θ over F. Show that if $\alpha \mid \beta$ in E, then $\alpha_i \mid \beta_i$ in E, where α_i and β_i are the conjugates of α and β respectively for $F(\theta)$.

3. Show that every algebraic integer divides a rational integer.

4. Find the norm of $1 + \sqrt{7}$ as an element of $R(\sqrt{7}, i)$.

5. Let θ be an integer in a field K. Suppose that θ and all its conjugates have absolute value (as complex numbers) less than 1. Prove that $\theta = 0$.

6. Tell whether $17 + 14\sqrt{7}$ is the fifth power of an integer in $R(\sqrt{7})$.

7. Let $(K/R) = 2$. Show that each rational prime p can be factored into at most two prime factors π_1 and π_2 of K.

8. Show that a conjugate of a unit is itself a unit.

9. Let θ be a root of the polynomial $x^3 + 3x + 7$. Show that θ is prime in $R(\theta)$.

10. Show that each integer α in a field can have at most $(\log \mid N\alpha \mid) / \log 2$ prime factors.

11. a. Factor $5 + \sqrt{3}$ in $R(\sqrt{3})$.
    b. Factor $7 + \sqrt{-5}$ in $R(\sqrt{-5})$.
    Hint. Factor the norm of each number.

12. Give an example of integers in a quadratic field which have equal norm but which are not conjugates or associates.

13. Let ϵ be an element of a quadratic field $R(\sqrt{D})$ satisfying $\epsilon\epsilon' = 1$, where $'$ denotes the conjugate. Show that $\epsilon = \gamma/\gamma'$, where γ is an integer of $R(\sqrt{D})$. Hint. Express ϵ as $a + b\sqrt{D}$, where $a, b \in R$, and find γ, γ' in terms of a and b.

14. Show that if α and β are integers in $R(\sqrt{D})$ with $N\alpha = N\beta$, then $\alpha/\beta = \gamma/\gamma'$, where γ is an integer of

$R(\sqrt{D})$. Give a non-trivial example.

Hint. Use the two preceding problems.

15. Show that if all elements of a quadratic field have non-negative norms, then the field is imaginary.

16. Prove that all units of $R(\sqrt{5})$ are of the form

$$\pm \{(1 + \sqrt{5})/2\}^n, \qquad n = 0, \pm 1, \pm 2, \ldots$$

17. Let K be a real field having a unit $u \neq \pm 1$. Show that there are units of K which are arbitrarily near zero (as real numbers).

18. a. Describe all the associates of $\sqrt{-3}$ in $R(\sqrt{-3})$.
    b. Describe all the associates of 2 in $R(\sqrt{2})$.

19. a. Show that 3 and $1 + \sqrt{-5}$ are relatively prime in $R(\sqrt{-5})$.
    b. Show that there exist no integers λ, μ in $R(\sqrt{-5})$ for which $3\lambda + (1 + \sqrt{-5})\mu = 1$.

20. Let I be an ideal in a field K and suppose that $1 \in I$. Show that I equals the ring of all integers in K. This ideal is usually denoted by (1).

21. Explain the difference between a basis and a set of generators of an ideal. Is a basis a set of generators, or vice versa?

22. a. Show that $\{4a + 2\sqrt{2}b \colon a, b \in J\}$ is an ideal in $R(\sqrt{2})$.
    b. Find a single integer that generates this ideal.
    c. What is a basis for this ideal?

23. Show that $\{5 + i, 2 + 3i\}$ is a basis for $(2 + 3i)$ in G.

24. Let $\omega_1, \ldots, \omega_n$ be an integral basis for a field K. Let α be a non-zero integer of K. Prove that $\alpha\omega_1, \ldots, \alpha\omega_n$ is a basis for an ideal I in K if and only if $I = (\alpha)$.

25. Let $I = (3 + i, 7 + i)$ be an ideal in $R(i)$.
    a. Find a single generator for I.
    b. Find a basis for I.

c. Draw a diagram showing the points of I in C, the complex numbers.

26. Consider I, the integer multiples of 4, as an ideal in each of the following fields. Give a basis for I and compute the discriminant of the basis. Compare it with the discriminant of the field.

   a. $R(i)$,   b. $R(\sqrt{-5})$,   c. $R(\sqrt{5})$.

27. Prove that every ideal in $R(i)$ is principal.

28. Let α and β be integers in a field K which have relatively prime norms. Show that the ideal $(\alpha, \beta) = (1)$.

29. Let θ be algebraic over F. Find a generator of the ideal

$$I = \{g(x) \in F[x]: g(\theta) = 0\}.$$

cf. Problem 4.4.

30. Let α be a root of $x^{17} - 43x^6 + 1$ and suppose $\alpha \in I$, an ideal in $R(\alpha)$. What can you say about I?

CHAPTER VIII

THE FUNDAMENTAL THEOREM OF IDEAL THEORY

1. **Basic properties of ideals.** According to the last chapter, every ideal in the algebraic number field K can be written $A = (\alpha_1, \ldots, \alpha_s)$. Under what circumstances can we say that A and $B = (\beta_1, \ldots, \beta_t)$ are the same ideal? The simple answer is given by

THEOREM 8.1. *The ideals* $A = (\alpha_1, \ldots, \alpha_s)$ *and* $B = (\beta_1, \ldots, \beta_t)$ *are the same if and only if each* α_i *can be written as*

$$\alpha_i = \sum_j \gamma_{ij}\beta_j$$

and each β_j *as*

$$\beta_j = \sum_i \delta_{ji}\alpha_i ,$$

where the $\gamma_{ij} , \delta_{ji}$ *are integers of* K.

The necessity of the condition is obvious. To prove the sufficiency let $\beta = \sum_j \lambda_j\beta_j$ be any element of B. Then $\beta = \sum_j \lambda_j \sum_i \delta_{ji}\alpha_i = \sum_i (\sum_j \lambda_j\delta_{ji})\alpha_i$, so that β is in A. Similarly every element of A is in B. Hence $A = B$.

COROLLARY 8.2. *Two non-zero principal ideals* (α) *and* (β) *are the same if and only if* α *and* β *are associated.*

If α and β are associated, $\alpha = \beta\epsilon$, where ϵ is a unit, and $\beta = \alpha(1/\epsilon)$, where $1/\epsilon$ is a unit. By the preceding theorem, $(\alpha) = (\beta)$. Conversely if $(\alpha) = (\beta)$ then $\alpha = \beta\gamma$, $\beta = \alpha\delta$, where γ and δ are integers in K. Hence $\alpha = (\alpha\delta)\gamma$, $1 = \delta\gamma$, so that $\alpha/\beta = \gamma$ is a unit.

By the *product* AB of the ideals $A = (\alpha_1, \ldots, \alpha_s)$ and $B = (\beta_1, \ldots, \beta_t)$ in K we mean the ideal

$$AB = (\alpha_1\beta_1, \ldots, \alpha_i\beta_j, \ldots, \alpha_s\beta_t)$$

in K generated by all products $\alpha_i\beta_j$. It is easily verified by means of Theorem 8.1 that the product AB is independent of the particular sets of generators chosen for the ideals A and B. It is a direct consequence of the definition of product that for any ideals A, B and C,

$$AB = BA$$

$$(AB)C = A(BC).$$

We shall say that A *divides* B, written $A \mid B$, if an ideal C exists so that $B = AC$. A is then called a *factor* of B. A *includes* B, written $A \supset B$, if every element of B is contained in A. A is then called a *divisor* of B. Note very carefully this distinction we are making between a factor and a divisor.

LEMMA 8.3. *If $A \mid B$ then $A \supset B$.*

In other words, a factor is a divisor. For suppose that $B = AC$, where $C = (\gamma_1, \ldots, \gamma_v)$. Then $(\beta_1, \ldots, \beta_t) = (\alpha_1\gamma_1, \ldots, \alpha_i\gamma_j, \ldots, \alpha_s\gamma_v)$, so every β_k is of the form $\sum_{i,j} \lambda_{ij}\alpha_i\gamma_j = \sum_i (\sum_j \lambda_{ij}\gamma_j)\alpha_i$ and is contained in A. Hence A includes B.

LEMMA 8.4. *A rational integer not zero belongs to at most a finite number of ideals in K.*

Let $\omega_1, \ldots, \omega_n$ be an integral basis for K. Then every integer of the field is of the form $\alpha = \sum_{i=1}^{n} c_i\omega_i$, where the c_i are rational integers.

Suppose a is a rational integer not zero and A an ideal containing it. Since $\pm a$ are both in A we can assume that

$a > 0$. Each c_i can be written

$$c_i = q_i a + r_i, \quad 0 \le r_i < a, 1 \le i \le n.$$

Then

$$\alpha = \sum (q_i a + r_i)\omega_i = a \sum q_i \omega_i + \sum r_i \omega_i = a\gamma + \beta,$$

where γ is an integer and β can take on only a finite number of different values (since $0 \le r_i < a$).

Let $A = (\alpha_1, \ldots, \alpha_s)$. Since $a \in A$, $A = (\alpha_1, \ldots, \alpha_s, a)$. Each α_i is of the form $a\gamma_i + \beta_i$ by the preceding remarks, so that $A = (a\gamma_1 + \beta_1, \ldots, a\gamma_s + \beta_s, a)$. By Theorem 8.1, $A = (\beta_1, \ldots, \beta_s, a)$. But β_i can take on only a finite number of different values for each $i = 1, \ldots, s$. So A must be one of only a finite number of ideals.

THEOREM 8.5. *An ideal $A \neq (0)$ has only a finite number of divisors.*

We first show that every non-zero ideal contains a non-zero rational integer. Let $\alpha \in A$, $\alpha \neq 0$. Let $\alpha = \alpha_1$, $\alpha_2, \ldots, \alpha_n$ be the conjugates of α for the field K. $\alpha_2, \ldots, \alpha_n$ are algebraic integers, but not necessarily in K. However,

$$\alpha_2 \cdots \alpha_n = N\alpha/\alpha \in K.$$

Thus $\alpha_2 \cdots \alpha_n$ is an integer of K and $N\alpha$ is in A.

Now we apply the preceding lemma to deduce that $N\alpha$ can belong to at most a finite number of ideals.

From Lemma 8.3 we have also the

COROLLARY 8.6. *An ideal $A \neq (0)$ has only a finite number of factors.*

It is our purpose to establish a theory of unique factorization for ideals similar to that obtained in Chapter I for the rational integers. The role of the units in the latter theory will be assumed by the ideal (1)—that is, the ring of all integers in K. The ideals which take over the function of the prime rational integers are naturally those ideals P which have no factors except P and (1). It is customary in the classical literature to call such ideals "prime", but in modern ring theory the word "prime" is reserved for another property of ideals which will be mentioned subsequently; so for the present we shall use instead the word "irreducible". Then an ideal P is *irreducible* if it has no factors except P and (1). What we shall eventually prove is that every ideal in K different from (0) and (1) can be represented as the product of irreducible ideals, uniquely to within order and to within multiplication by (1).

We shall give two proofs of this important theorem: a modification of a classical proof based on ideas of A. Hurwitz and a more modern proof due to E. Noether and W. Krull. These two proofs will be given in the following two sections, which can be read independently of one another.

It is useful to introduce two further kinds of ideals which will eventually turn out to be equivalent to one another and to irreducible ideals. An ideal A is *maximal* if it has no divisors except (1) and A—that is, if it is included in no larger ideal except (1). An ideal P different from (0) or (1) is *prime* if it has the following property: *whenever a product of integers $\gamma\delta$ is in P, so is either γ or δ.*

THEOREM 8.7. *An ideal P different from (0) or (1) is maximal if and only if it is prime.*

First suppose that $P = (\alpha_1, \ldots, \alpha_s)$ is maximal and let it contain $\gamma\delta$. If it contains γ we are through. Suppose it does not contain γ; we shall show it contains δ. Let $P' = (\alpha_1, \ldots, \alpha_s, \gamma)$. Then $P' \supset P$. But P is maximal, so $P' = P$ or $P' = (1)$. $P' = P$ is impossible, for then γ belongs to P. Hence $P' = (1)$, so that 1 is contained in P'.

1 can therefore be written in the form

$$1 = \lambda_1\alpha_1 + \cdots + \lambda_s\alpha_s + \lambda\gamma,$$

where $\lambda_1, \ldots, \lambda_s, \lambda$ are integers of K, so that

$$\delta = (\lambda_1\delta)\alpha_1 + \cdots + (\lambda_s\delta)\alpha_s + \lambda(\gamma\delta).$$

Since $\alpha_1, \ldots, \alpha_s$ and $\gamma\delta$ are in P, so is δ. Hence P is prime.

Conversely, let $P = (\alpha_1, \ldots, \alpha_s)$ be a prime ideal. Let $P' \supset P$, $P' \neq P$. We must show that $P' = (1)$. Let α be an integer in P' but not in P. Form its powers α^j; they are in P'.

Let $\omega_1, \ldots, \omega_n$ be an integral basis for K. Let $\beta \neq 0$ be any integer in P. Then $\pm N\beta$ are in P, so that P contains a positive rational integer a. According to the proof of Lemma 8.4 each integer in K can be written in the form $a\gamma + \sum_{i=1}^{n} r_i\omega_i$, where each of the r_i can take on only a finite number of different values. In particular each α^j is of the form

$$\alpha^j = a\gamma_j + \sum_{i=1}^{n} r_{ij}\omega_i.$$

Then $\alpha^j - a\gamma_j$ can take on only a finite number of different values. So there is a pair of integers $k, l, k > l$, such that

$$\alpha^k - a\gamma_k = \alpha^l - a\gamma_l.$$

$\alpha^k - \alpha^l = a(\gamma_k - \gamma_l)$ is in P, by the choice of a. Then $\alpha^l(\alpha^{k-l} - 1)$ is in P. Since P is a prime ideal, one of the two factors α^l and $\alpha^{k-l} - 1$ must be in P.

Now $\alpha^l = \alpha \cdot \alpha \cdots \alpha$ cannot be in P, for otherwise one of the factors α would be, and α was chosen as an integer not in P. Hence $\alpha^{k-l} - 1$ is in P. Since $P' \supset P$, $\alpha^{k-l} - 1$ is in P'. But every power of α is in P', α^{k-l} in particular. Hence -1 belongs to P', so that $P' = (1)$. It follows that P is maximal.

COROLLARY 8.8. *If P is a maximal ideal and $P \supset AB$, then $P \supset A$ or $P \supset B$.*

If $P \supset A$ we are done. Suppose α is in A but does not belong to P. If β is in B then $\alpha\beta$ is in P, for $P \supset AB$. But P is prime, according to Theorem 8.7, so that P contains β. Hence every element of B is contained in P, $P \supset B$.

2. The classical proof of the unique factorization theorem. We begin with the following lemma.

LEMMA 8.9. *Every ideal A different from* (0) *and* (1) *has a maximal divisor.*

By Theorem 8.5 the ideal A has only a finite number of divisors. Any divisor B of A, $B \neq A$, has fewer divisors than A, for any divisor of B is a divisor of A, since $B \supset A$, and moreover A has a divisor which B does not, namely A itself.

Among the divisors of A choose one different from (1) with the smallest number of divisors. This is possible, by Theorem 8.5. Call it P. Then P is maximal. If it were not, then there would be an ideal $P' \neq$ (1) such that $P' \supset P$, $P' \neq P$. But then P' has fewer divisors than P, and $P' \supset A$, contrary to the choice of P.

The following lemmas will be used to establish the converse of Lemma 8.3.

LEMMA 8.10. *If*

$$f(x) = \delta_m x^m + \delta_{m-1} x^{m-1} + \cdots + \delta_0 \quad (\delta_m \neq 0)$$

is a polynomial with all its coefficients algebraic integers and ρ is one of its roots, then all the coefficients of the polynomial $f(x)/(x - \rho)$ are algebraic integers.

By Theorem 6.4, $\delta_m \rho$ is an algebraic integer, for it satisfies the equation

$$x^m + \delta_{m-1}x^{m-1} + \cdots + \delta_m^{m-2}\delta_1 x + \delta_m^{m-1}\delta_0 = 0.$$

The lemma is certainly true if $m = 1$. Suppose it has been established for all polynomials of degree $\leq m - 1$. Since

$$\phi(x) = f(x) - \delta_m x^{m-1}(x - \rho)$$

is of degree $\leq m - 1$, and since $\phi(\rho) = 0$, the polynomial

$$\frac{\phi(x)}{x - \rho} = \frac{f(x)}{x - \rho} - \delta_m x^{m-1}$$

has integral coefficients. Then so has $f(x)/(x - \rho)$. This completes the induction.

LEMMA 8.11. *If $f(x)$ is the polynomial of Lemma 8.10 and*

$$f(x) = \delta_m(x - \rho_1) \cdots (x - \rho_m),$$

then $\delta_m \rho_1 \cdots \rho_k$ is an algebraic integer for $k = 1, 2, \ldots m$.

For by successive applications of the preceding lemma,

$$\frac{f(x)}{(x - \rho_{k+1}) \cdots (x - \rho_m)} = \delta_m(x - \rho_1) \cdots (x - \rho_k)$$

has only integral coefficients.

The next lemma is a generalization of that of Gauss (Theorem 3.6).

LEMMA 8.12. *Let*

$$p(x) = \alpha_p x^p + \alpha_{p-1}x^{p-1} + \cdots + \alpha_0,$$

$$q(x) = \beta_q x^q + \beta_{q-1}x^{q-1} + \cdots + \beta_0$$

be polynomials with integral coefficients, $\alpha_p \beta_q \neq 0$. Let

$$r(x) = p(x)q(x) = \gamma_r x^r + \gamma_{r-1}x^{r-1} + \cdots + \gamma_0.$$

If δ is an integer such that all γ_k/δ are integers, so are all $\alpha_i\beta_j/\delta$.

For suppose

$$p(x) = \alpha_p(x - \rho_1) \cdots (x - \rho_p),$$

$$q(x) = \beta_q(x - \omega_1) \cdots (x - \omega_q);$$

then

$$\frac{r(x)}{\delta} = \frac{\alpha_p\beta_q}{\delta} (x - \rho_1) \cdots (x - \rho_p)(x - \omega_1) \cdots (x - \omega_q)$$

has integral coefficients. By Lemma 8.11 every product

$$(8.1) \qquad \frac{\alpha_p\beta_q}{\delta} \rho_{m_1} \cdots \rho_{m_i}\omega_{n_1} \cdots \omega_{n_j}$$

is an integer. But α_i/α_p and β_j/β_q are elementary symmetric functions in the ρ's and ω's respectively, so that

$$\frac{\alpha_i\beta_j}{\delta} = \frac{\alpha_p\beta_q}{\delta} \frac{\alpha_i}{\alpha_p} \frac{\beta_j}{\beta_q}$$

is a sum of terms of the form (8.1). Hence $\alpha_i\beta_j/\delta$ is an algebraic integer.

THEOREM 8.13. *For every ideal $A \neq (0)$ there is an ideal $B \neq (0)$ such that AB is principal. In fact we can find B so that $AB = (a)$, where a is a rational integer.*

Let $A = (\alpha_1, \ldots, \alpha_r)$ and define

$$g_{i_1, \ldots, i_r}(x) = \alpha_1^{(i_1)}x + \alpha_2^{(i_2)}x^2 + \ldots + \alpha_r^{(i_r)}x^r,$$

where $1 \leq i_1 \leq n, \ldots, 1 \leq i_r \leq n$ and the $\alpha_j^{(i_j)}$, $i_j = 1, \ldots, n$, are the conjugates of a_j for K. Form the polynomial

$$F(x) = \prod_{i_1,\ldots,i_r=1}^{n} g_{i_1, \ldots, i_r}(x) = \sum c_p x^p.$$

The argument used in proving Theorem 6.4 shows that each coefficient c_p is a rational integer. Let $g_1(x) = g_{1,1,\ldots,1}(x)$ be the polynomial having the original a_i as coefficients. Now $g_1(x) \mid F(x)$, and the quotient

$$h(x) = F(x) / g_1(x) = \beta_1 x + \ldots + \beta_m x^m$$

has coefficients which are integers in K.

Let a be the greatest common divisor of the c_p so that $F(x)/a$ is primitive. Define $B = (\beta_1, \ldots, \beta_m)$. We shall show that $AB = (a)$.

By Lemma 8.12 a divides all $\alpha_i\beta_j$. But AB is generated by all the products $\alpha_i\beta_j$. Hence $(a) \supset AB$. On the other hand, since a is the greatest common factor of the c_p, the rational integers c_k/a are relatively prime. Then there exist integers x_k such that*

$$1 = \sum x_k \frac{c_k}{a}, \qquad a = \sum x_k c_k.$$

But each c_k is, by its definition, of the form $\sum \lambda_{ijk}\alpha_i\beta_j$ with the $\lambda_{ijk} = 0$ or 1, so a is of the form $\sum_{i,j} \left(\sum_k x_k\lambda_{ijk} \right) \alpha_i\beta_j$. Then a is in AB, $AB \supset (a)$. So finally $(a) = AB$.

COROLLARY 8.14. *If* $AB = AC, A \neq (0)$, *then* $B = C$.

For let $AD = (\delta)$, a principal ideal. Then $ABD = ACD$, $(\delta)B = (\delta)C$. Then δ times each integer in B equals δ times some integer in C, so each integer of B is in C, $C \supset B$. Similarly $B \supset C$, so that $B = C$.

COROLLARY 8.15. (*Converse of Lemma 8.3*). *If* $A \supset B$, *then* $A \mid B$. *In other words, a divisor is a factor.*

---

* This can be proved in the same manner as Theorem 1.2.

Choose D so that $AD = (\delta)$. Since $A \supset B$, $AD \supset BD$; this follows from the definition of multiplication of ideals. Write $BD = (\rho_1, \ldots, \rho_m)$. Each ρ_i is contained in $AD = (\delta)$, and is therefore of the form $\lambda_i \delta$. Hence

$$BD = (\delta)(\lambda_1, \lambda_2, \ldots, \lambda_m) = AD(\lambda_1, \lambda_2, \ldots, \lambda_m).$$

By Corollary 8.14, $B = A(\lambda_1, \lambda_2, \ldots, \lambda_m)$, so $A \mid B$.

COROLLARY 8.16. *An ideal is maximal if and only if it is irreducible.*

For it has now been established that factors and divisors are the same, so that an ideal which lacks one lacks the other.

LEMMA 8.17. *If $B \mid A$ and $B \neq A$, then B has fewer factors than A.*

This has already been proved in the course of establishing Lemma 8.9, since divisors and factors are now known to be the same thing.

LEMMA 8.18. *Every ideal not (0) or (1) can be factored into the product of irreducible ideals.*

By Lemma 8.9 the ideal A has a maximal divisor P_1, which by virtue of Corollary 8.15 is also a factor. Then $A = P_1 A_1$. If A_1 is (1) or maximal, stop here. Otherwise repeat the procedure with $A_1 = P_2 A_2$ to obtain $A = P_1 P_2 A_2$, and continue in this way. Eventually the procedure must stop, since each of A_1, A_2, $\ldots$ divides its predecessor, and so by Lemma 8.17 has fewer factors than the predecessor. We can conclude that $A = P_1 P_2 \cdots P_r$, where each P_i is maximal. By Corollary 8.16 each P_i is irreducible, and the lemma is proved.

To prove the uniqueness of the factorization we shall use the following consequence of Corollaries 8.8, 8.15, and 8.16.

LEMMA 8.19. *If P is an irreducible ideal and $P \mid AB$, then $P \mid A$ or $P \mid B$.*

THEOREM 8.20. (*The Fundamental Theorem*). *Every ideal not (0) or (1) can be factored into the product of irreducible ideals. This factorization is unique except for the order of the factors.*

The first part of the theorem has already been established as Lemma 8.18. We turn to the uniqueness. Suppose that the ideal A has two factorizations into irreducible ideals distinct from (1):

$$(8.2) \qquad A = P_1 P_2 \cdots P_r = P_1' P_2' \cdots P_s', \qquad s \geq r.$$

By Lemma 8.19 the ideal P_1' must divide one of the P_i, say P_1. Then $P_1' \mid P_1$, $P_1' \supset P_1$. But P_1 is maximal, and $P_1' \neq (1)$, so that $P_1' = P_1$. By Corollary 8.14 we can divide out P_1 in (8.2) to obtain

$$P_2 \cdots P_r = P_2' \cdots P_s'.$$

We can repeat this procedure until all the factors on the left-hand side are exhausted. Suppose there remains a factor P_i' on the right-hand side. Then $P_i' \mid (1)$, $P_i' \supset (1)$, $P_i' = (1)$. Hence all the factors on the right-hand side are also used up, so that, after some rearrangement of the order of the P_i if necessary, $P_i = P_i'$, $i = 1, \ldots, r$, and $r = s$. The proof is complete.

3. The modern proof. The following version follows the outline given by Ore in his survey (see the bibliography). *We shall not make any use of the results obtained in* §2. Instead we shall proceed directly to a proof of a modified form of Theorem 8.20 in which "irreducible" is replaced by "maximal". As a *consequence* of this we shall then establish Corollaries 8.15 and 8.16. This will enable us to restore the word "irreducible" for "maximal", and thus prove Theorem 8.20 in its final form. The reader is reminded that the last theorem we are at liberty to use in this paragraph is Theorem 8.7 which asserts the equivalence of maximal and prime ideals. The equivalence of factors and divisors is established in §2, and this fact we cannot use without first offering a proof.

LEMMA 8.21. *An ideal A not (0) or (1) is a divisor of a product $P_1 \cdots P_s$ where each P_i is a divisor of A, and is a maximal ideal.*

If A is maximal there is nothing to prove. If it is not maximal, then by Theorem 8.7 A contains a product $\beta\gamma$ such that neither β nor γ belong to it. If $A = (\alpha_1, \ldots, \alpha_r)$, let

$$B = (\alpha_1, \ldots, \alpha_r, \beta), \qquad C = (\alpha_1, \ldots, \alpha_r, \gamma).$$

Then $A \supset BC$, $B \supset A$, $C \supset A$. Now repeat the procedure with B and C, and continue. At each stage the new ideals all include A and their product is included in A. But the procedure must stop, by Theorem 8.5, so that we finally reach maximal ideals.

Let P be a maximal ideal. We define P^{-1} as the totality of numbers α in the field K, *integers or not*, such that the product $\alpha\pi$ is an integer (not necessarily in P) for all numbers π in P.

LEMMA 8.22. *If P is a maximal ideal and $P \neq (1)$, then P^{-1} contains a number which is not an algebraic integer.*

Let π, different from zero, be an integer in P and consider the principal ideal (π). (π) includes a product $P_1 \cdots P_r$ of maximal ideals, by the preceding lemma. If there are several such products, pick one for which r is least. Now $P \supset (\pi) \supset P_1 \cdots P_r$, so, by Corollary 8.8, P contains one of the P_i, say P_1. Since P_1 is maximal, $P = P_1$. The ideal (π) does not include $P_2 \cdots P_r$ since the product with the least number of factors was picked to begin with. Then $P_2 \cdots P_r$ contains an integer γ not in (π). Consequently γ/π is not an integer but $(\pi) \supset P P_2 \cdots P_r \supset P(\gamma)$. This means that if π' is in P then $\pi'\gamma$ is in (π). Then $\pi'\gamma/\pi$ is an integer. Hence γ/π is in P^{-1}.

If A is an ideal we define the product $AP^{-1} = P^{-1}A$ to be the set of all finite sums of products $\alpha\beta$, where α is in A and β in P^{-1}. We remark that a set AP^{-1} is an example of a generalization of ideals called fractional ideals. Note that if $P \supset A$, $\alpha \in A$, and $\beta \in P^{-1}$, then $\alpha\beta$ is an integer, and it is easy to see that AP^{-1} is an ideal.

LEMMA 8.23. *If P is a maximal ideal, then $PP^{-1} = (1)$.*

Let $A = PP^{-1}$. A is an ideal (why?). Since P^{-1} contains 1, $A \supset P$. But P is maximal, so that $A = (1)$ or $A = P$.

We assume that $A = P$. This will lead to a contradiction.

Let $\omega_1, \ldots, \omega_n$ be a basis for P, and let $\gamma_1 = \gamma/\pi$ be a non-integer in P^{-1} (see the preceding lemma). The products $\gamma_1\omega_i$ are all in $A = P$ and so can be represented as

$$\gamma_1\omega_i = \sum_{j=1}^{n} a_{ij}\omega_j,$$

where the a_{ij} are rational integers. Then the system of equations

$$(a_{11} - \gamma_1)x_1 + a_{12}x_2 + \cdots = 0$$

$$a_{21}x_1 + (a_{22} - \gamma_1)x_2 + \cdots = 0$$

$$\cdots\cdots\cdots\cdots\cdots\cdots\cdots\cdots\cdots\cdots$$

$$a_{n1}x_1 + \cdots + (a_{nn} - \gamma_1)x_n = 0$$

has a non-trivial solution $x_i = \omega_i$, so the determinant

$$\begin{vmatrix} a_{11} - \gamma_1 & a_{12} & \cdots & \cdots \\ a_{21} & a_{22} - \gamma_1 & \cdots & \cdots \\ \cdots\cdots\cdots\cdots & \cdots\cdots\cdots\cdots & \cdots & \cdots \\ a_{n1} & a_{n2} & \cdots & a_{nn} - \gamma_1 \end{vmatrix}$$

vanishes. Hence γ_1 satisfies a monic equation with integral coefficients and is therefore an algebraic integer. This contradiction leads to the conclusion $A = (1)$.

LEMMA 8.24. *Every ideal A not* (0) *or* (1) *is the product of maximal ideals.*

By Lemma 8.21 A includes a product $P_1 \cdots P_r$ of maximal ideals where each P_i is a divisor of A, and as before we choose the product for which r is least. We proceed by induction on r.

If A includes only one maximal P, then $A = P$ and we are done. Suppose the theorem is valid for ideals which include a product of fewer than r factors. Since $A \supset P_1 \cdots P_r$, then $AP_r^{-1} \supset P_1 \cdots P_{r-1}$, by Lemma 8.23. AP_r^{-1} is an ideal because $P_r \supset A$. By the hypothesis of the induction AP_r^{-1} is a product $P_1' P_2' \cdots P_k'$ of maximal ideals. By Lemma 8.23 once more $A = P_1' P_2' \cdots P_k' P_r$, so that A is a product of maximal ideals.

LEMMA 8.25. *Let* $A = P_1 \cdots P_r$ *and* $B = Q_1 \cdots Q_s$ *be products of maximal ideals each* $\neq (1)$. *If* $B \supset A$, *then each ideal* Q_i *occurs among the* P_j *at least as many times as it occurs in B.*

Since Q_1 is a factor of B it is a divisor of B (Lemma 8.3). Hence $Q_1 \supset B \supset A = P_1 \cdots P_r$. By Corollary 8.8, Q_1 contains one of the P_i, say P_1; so $Q_1 = P_1$, each being maximal and $\neq (1)$. Also

$$P_1^{-1}B \supset P_1^{-1}A = P_2 \cdots P_r,$$

by Lemma 8.23. The result then follows by induction if we assume it to be true when B contains fewer than r factors.

LEMMA 8.26. *The representation of an ideal as the product of maximal ideals is unique to within order.*

For suppose

$$A = P_1 P_2 \cdots P_r = Q_1 Q_2 \cdots Q_s .$$

Then we need only apply the preceding lemma with $A = B$. As a result of Lemma 8.24 and 8.26 we have

THEOREM 8.27. *An ideal different from* (0) *and* (1) *can be represented, uniquely apart from order, as the product of maximal ideals.*

In order to prove the fundamental Theorem 8.20 it is enough to show that the word "maximal" in the preceding theorem can be replaced by "irreducible". This is justifiable if we can prove that a divisor is a factor—in other words: if $B \supset A$, then $B \mid A$. But this in turn follows from Lemma 8.25. For we may write $A = P_1^{e_1} \cdots P_r^{e_r}$ and $B = P_1^{f_1} \cdots P_r^{f_r}$, where the P_i are the distinct maximal factors of A and B, and $e_i \geq f_i$. So $A = BC$, where $C = P_1^{e_1-f_1} \cdots P_r^{e_r-f_r}$. Hence the fundamental theorem is established.

If we exclude from consideration (0) and (1), then maximal, irreducible, and prime ideals are the same. This follows from Theorem 8.7 and Corollary 8.16. The use of a result from §8.2 can be avoided by using Lemma 8.3 and the fact established just above that a divisor is a factor. The literature on algebraic numbers uses the term "prime" most frequently, and in the sequel we shall adhere to that tradition.

Problems

1. Prove that the ideals $(3, 1 + 2\sqrt{-5})$ and $(7, 1 - 2\sqrt{-5})$ in $R(\sqrt{-5})$ are not equal by showing that the representation

$$3 = 7(a + b\sqrt{-5}) + (1 - 2\sqrt{-5})(c + d\sqrt{-5})$$

is impossible for $a, b, c, d \in J$.

2. State a necessary and sufficient condition, based on Theorem 8.1, for one ideal to contain another.

3. Suppose that

$$A = (a_1, \ldots, a_s) = (\alpha_1, \ldots, \alpha_\sigma),$$
$$B = (b_1, \ldots, b_t) = (\beta_1, \ldots, \beta_\tau).$$

Prove that AB is well defined by showing that

$$(a_1b_1, \ldots, a_ib_j, \ldots, a_sb_t)$$
$$= (\alpha_1\beta_1, \ldots, \alpha_i\beta_j, \ldots, \alpha_\sigma\beta_\tau).$$

4. Prove that the following is an equivalent definition of the product of ideals A and B:

$$AB = \{\sum a_ib_i : a_i \in A, b_i \in B\}.$$

(Each sum extends over a finite number of terms of A and B.)

5. Prove the following assertions about ideals:

(a) $\alpha \in A, \beta \in B \Rightarrow \alpha\beta \in AB$ (c) $AB = BA$

(b) $(AB)C = A(BC)$ (d) $A \supset AB$

6. Show that in $R(\sqrt{-5})$ we have

(a) $(3, 1 + 2\sqrt{-5})(3, 1 - 2\sqrt{-5}) = (3)$

(b) $(7, 1 + 2\sqrt{-5})(7, 1 - 2\sqrt{-5}) = (7)$.

7. (a) Find two factorizations of 10 in $R(\sqrt{-6})$.

(b) Factor (10) into four factors in $R(\sqrt{-6})$.

Hint. Use the analogy with the factorization of (21) in $R(\sqrt{-5})$.

8. Let α, β be integers in some field. Show that if $(\alpha) \mid (\beta)$, then $\alpha \mid \beta$.

9. If A and B are ideals and $A \mid B$ and $B \mid A$, show that $A = B$. Characterize the ideals which divide (1).

10. Let $A \neq (0)$ and B be ideals and suppose $AB = A$. Using only the definition of product of ideals and the existence of a basis for an ideal, prove that $B = (1)$. Hint. Let $\omega_1, \ldots, \omega_n$ be a basis for A. Then $\omega_1 \in A = AB$ and so $\omega_1 = \sum_{i=1}^{r} a_i b_i$. Express the a_i, $1 \leq i \leq r$, in terms of the basis and deduce that $1 \in B$.

11. Every non-zero ideal contains an infinite number of rational integers.

12. Let $K \supset R$ and $(K/R) = n$. Show that a rational integer $b \neq 0$ can belong to at most b^n ideals of K.

13. Show that $I = (1 + \sqrt{-5}, 1 - \sqrt{-5})$ is a maximal ideal in $R(\sqrt{-5})$. Make a sketch of I in the complex plane.

14. Prove directly that a maximal ideal is irreducible.

15. Let π be a prime in an algebraic number field K. Tell whether (π) is necessarily a prime ideal in K.

16. Suppose that a and $b \in J$ and (a, b) is a maximal ideal in R. What can you say about a and b?

17. Show that $(3, 1 + 2\sqrt{-5}) \mid (1 + 2\sqrt{-5})$ in $R(\sqrt{-5})$. Using the method described in the proof of Theorem 8.13 and Corollary 8.15 find an ideal B in $R(\sqrt{-5})$ satisfying $(3, 1 + 2\sqrt{-5})B = (1 + 2\sqrt{-5})$.

18. Let $\alpha \in I$, an ideal of a field K. Show that there exists an ideal I' in K such that $(\alpha) = I'I$.

19. (a) Find $(2)^{-1}$ in R.

   (b) Find $(1 + i)^{-1}$ in $R(i)$.

   (c) Find $(3, 1 + 2\sqrt{-5})^{-1}$ in $R(\sqrt{-5})$.

20. Let P be a maximal ideal in a field K. Prove that

   (a) if λ_1, λ_2 are integers of K and β_1, $\beta_2 \in P^{-1}$, then

$\lambda_1\beta_1 + \lambda_2\beta_2 \in P^{-1}$;

(b) if $P \supset A$, then AP^{-1} is an ideal;

(c) if $P \supset A \supset B$, then $AP^{-1} \supset BP^{-1}$.

21. Let $A \neq (0)$ be an ideal of a field K. Let $\beta \in K$ and have the property that $\beta a \in A$ for each $a \in A$. Prove that β is an integer of K.

CHAPTER IX

CONSEQUENCES OF THE FUNDAMENTAL THEOREM

1. **The highest common factor of two ideals.** Let A and B be two ideals in the algebraic number field K. An ideal C is said to be a *highest common factor of A and B*, written (A, B), if $C \mid A$ and $C \mid B$, and if every ideal which divides both A and B divides C. A highest common factor is unique, for suppose both C and D have the requisite properties. Then $C \mid D$ and $D \mid C$. By Lemma 8.3, $C \supset D$ and $D \supset C$, so that $C = D$.

There is a simple way of obtaining (A, B), as follows. Let $A = (\alpha_1, \ldots, \alpha_r)$, $B = (\beta_1, \ldots, \beta_s)$. Define $D = (\alpha_1, \ldots, \alpha_r, \beta_1, \ldots, \beta_s)$. Then $D = (A, B)$. For clearly $D \supset A$, $D \supset B$ so that (Corollary 8.15) $D \mid A$, $D \mid B$. Further suppose $E \mid A$, $E \mid B$. Then $E \supset A$, $E \supset B$, so that $E \supset D$, hence $E \mid D$. Still another method of obtaining the highest common factor of A and B is this: Let $P_1, \ldots, P_r$ be the totality of distinct prime ideals which occur in the factorizations of both A and B. Then $(A, B) = P_1^{e_1} \cdots P_r^{e_r}$, where e_i is the highest power (possibly zero*) for which $P_i^{e_i}$ divides both A and B. This establishes

THEOREM 9.1. *Two ideals A and B, not both 0, have a unique highest common factor (A, B).*

If $(A, B) = (1)$ we say that A and B are *relatively*

---

* It is convenient to define the power C^0 of an ideal C as (1).

prime. It is customary in this case to write simply $(A, B) = 1$.

We saw earlier (Theorem 7.11) that not every ideal in a field K need be principal. We are now in a position to show that in any case an ideal can always be generated by *two* elements of K.

LEMMA 9.2. *If A and B are ideals different from* (0), *there is an integer α in A such that*

$$\left(\frac{(\alpha)}{A}, B\right) = 1.$$

(If α is in A, then $A \supset (\alpha)$, $A \mid (\alpha)$, so we may define the ideal $(\alpha)/A$ to be the solution of the equation $AX = (\alpha)$.)

If $B = (1)$ then the lemma is trivial, for there we can take for α any element of A. So we suppose that $B \neq (1)$.

Let $P_1, \ldots, P_r$ be the distinct prime factors of B. If $r = 1$ then $B = P^j$, $j > 0$, so we need only find an α in A for which

(9.1) $$((\alpha)/A, P) = 1.$$

We shall use repeatedly the fact that (9.1) is equivalent to

(9.2) $$\alpha \in A \quad \text{and} \quad \alpha \notin AP.$$

Indeed, if (9.2) holds, then $A \supset (\alpha)$ and $A \mid (\alpha)$, so that $(\alpha) = AC$ for some ideal C. If $(C, P) \neq 1$, then C and P have the highest common factor P. Consequently $C = PD$, $(\alpha) = APD$, $AP \mid (\alpha)$, and $AP \supset (\alpha)$, contrary to (9.2). Thus (9.1) must hold. Conversely, if (9.1) holds, then $((\alpha), AP) = A$, $A \mid (\alpha)$, $A \supset (\alpha)$ and $\alpha \in A$. Further, if $\alpha \in AP$, then $(\alpha) \subset AP$, $AP \mid (\alpha)$, and $A = ((\alpha), AP) = AP$. This is impossible, since $A \neq 0$ and $P \neq (1)$. Thus (9.2) holds.

There exists an integer α satisfying (9.2), for otherwise $AP \supset A$, $AP \mid A$, and $P \mid (1)$. This completes the case $r = 1$.

For $r > 1$, it suffices to find an element α for which (9.1) holds for each of the cases $P = P_1, \ldots, P_r$. For $m = 1, \ldots, r$, consider ideals $A_m = AP_1 \cdots P_r/P_m$ and P_m. The preceding argument shows that there exists an element $\alpha_m \in A_m$ such that (9.1) holds with α_m, A_m, and P_m in place of α, A, and P respectively. We set $\alpha = \alpha_1 + \cdots + \alpha_r$. Now $A \supset A_m$ since $A \mid A_m$, and hence $\alpha_m \in A$ for $m = 1, \ldots, r$. Thus $\alpha \in A$.

We shall show that $\alpha \notin AP_m$ and then conclude that (9.1) holds for $P = P_m$. For $i \neq m$, $\alpha_i \in AP_m$ since

$$(\alpha_i) \subset A_i = \frac{AP_1 \cdots P_r}{P_i} = AP_m \frac{P_1 \cdots P_r}{P_i P_m} \subset AP_m.$$

The preceding relation also implies that $AP_i \mid (\alpha_m)$ or $P_i \mid (\alpha_m)/A$ for all $i \neq m$. If $\alpha_m \in AP_m$, then $P_m \mid (\alpha_m)/A$ and hence $P_1 P_2 \cdots P_m \mid (\alpha_m)/A$ or $\alpha \in AP_1 \cdots P_m = A_m P_m$. This would contradict (9.2) for α_m, A_m, and P_m. Thus $\alpha_m \notin AP_m$, and since the other $\alpha_i \in AP_m$, we have $\alpha \notin AP_m$. It follows that (9.1) holds for $P = P_m$.

THEOREM 9.3. *Let A be an ideal not zero, and β any non-zero element in it. Then we can find an α in A such that $A = (\alpha, \beta)$.*

Define $B = (\beta)/A$. By the preceding lemma there is an α in A such that

$$\left(\frac{(\alpha)}{A}, B \right) = \left(\frac{(\alpha)}{A}, \frac{(\beta)}{A} \right) = 1.$$

Let $(\alpha) = AC$, $C = (\alpha)/A$. Since $(\beta) = AB$ and $(B, C) = 1$, the highest common factor of (α) and (β) is A. By the remarks preceding Theorem 9.1 $A = (\alpha, \beta)$.

Observe that we have made frequent use of the quotient $(\alpha)/A$ when α is in A. In the future we shall write this as α/A and understand $A \mid \alpha$ to mean $A \mid (\alpha)$. α is in A if and only if $A \mid \alpha$. Another notation is $\alpha \equiv 0 \pmod{A}$ or $\alpha \equiv 0(A)$.

2. Unique factorization of integers.

We return now to the problem of unique factorization of integers in K, a question temporarily abandoned in Chapter VII. Our next theorem confirms a conjecture made there.

THEOREM 9.4. *The factorization of integers of K into primes is unique (to within order and units) if and only if all the ideals in K are principal.*

That such a factorization is possible has already been settled by Theorem 7.5.

First assume that all the ideals in K are principal. Suppose an element of K, not zero or a unit, has two factorizations into prime integers:

$$\alpha = \pi_1 \cdots \pi_s = \pi_1' \cdots \pi_t'.$$

Clearly

$$(9.3) \qquad (\alpha) = (\pi_1) \cdots (\pi_s) = (\pi_1') \cdots (\pi_t').$$

If π is a prime integer, then (π) is a prime ideal. For suppose $(\pi) = BC$. Since B and C are both principal by hypothesis, $(\pi) = (\beta)(\gamma) = (\beta\gamma)$. By Corollary 8.2, π and $\beta\gamma$ are associated, so one of β or γ is a unit. Hence one of B and C is the ideal (1), and (π) is prime. Then (9.3) gives two factorizations of (α) into prime ideals. By the uniqueness of factorization of ideals we must have $s = t$, and $(\pi_i) = (\pi_i')$ after a suitable rearrangement of factors. Moreover, π_i/π_i' is a unit. This proves the sufficiency.

Suppose conversely, that factorization of integers is

unique. To prove that every ideal is principal it is enough to prove that every prime ideal P is principal. Let a be a non-zero integer in P. Then $P \mid a$. Let $a = \pi_1 \cdots \pi_r$ be the factorization of a into prime integers in K. Then $(a) = (\pi_1) \cdots (\pi_r)$ so that $P \mid \pi$ for some prime π in K. Now (π) is a prime ideal, for if $cd \in (\pi)$, then $\pi \mid cd$ and by the hypothesis of unique factorization $\pi \mid c$ or $\pi \mid d$, i.e., $c \in (\pi)$ or $d \in (\pi)$. Since P is prime, $P = (\pi)$.

We shall now present a criterion for the principality of all ideals in K—that is, for uniqueness of factorization of integers. The criterion is a generalization of the division algorithm, Theorem 1.1, which we used in proving the corresponding result for integers in R. It is due to Dedekind and Hasse.

THEOREM 9.5. *Every ideal in K is principal if, and only if, for every two integers α and β, neither zero, such that $\beta \nmid \alpha$ and $\mid N\alpha \mid \geq \mid N\beta \mid$, there exist integers γ and δ such that*

$$0 < \mid N(\alpha\gamma - \beta\delta) \mid < \mid N\beta \mid.$$

First suppose that every ideal in K is principal. Let α, β be integers of the prescribed kind and let $A = (\alpha, \beta)$. Since A is principal $(\alpha, \beta) = (\omega)$, so every integer in A is a multiple of ω. In particular $\beta = \sigma\omega$, $N\beta = N\sigma N\omega$. β and ω are not associated, for otherwise $\beta \mid \alpha$ since we know that $\omega \mid \alpha$. Hence $\mid N\sigma \mid > 1$, so that $\mid N\omega \mid < \mid N\beta \mid$. But ω is in (α, β), so $\omega = \alpha\gamma - \beta\delta$, and therefore $\mid N(\alpha\gamma - \beta\delta) \mid < \mid N\beta \mid$. Finally $\omega \neq 0$, since $\beta = \sigma\omega$, so that $\mid N(\omega) \mid > 0$.

Conversely, suppose the criterion to be satisfied and let A be any non-zero ideal in K. By Theorem 9.3 we can write it as $A = (\alpha, \beta)$. Let ω be a non-zero element of A for which $\mid N(\omega) \mid$ is least. Then $A = (\omega)$, for if γ is any integer in A such that $\omega \nmid \gamma$ we can find a combination

$\mu\gamma - \nu\omega$ in A such that

$$0 < |N(\mu\gamma - \nu\omega)| < |N\omega|.$$

This contradicts the choice of ω, so there can be no γ in A for which $\omega \nmid \gamma$.

The criterion just established is unfortunately very difficult to apply in practice. Sometimes it is possible to apply it with $\gamma = 1$. In this case the field is called *Euclidean*. The number of real Euclidean quadratic fields is finite. (See Hardy and Wright, Chapter XIV, and Stark, Chapter VIII, for a discussion of this and the following remarks.) The only imaginary quadratic fields $R(\sqrt{D})$, D square-free, which are Euclidean are those for which $D = -1, -2, -3, -7, -11$. In addition, the following satisfy the more general Dedekind-Hasse criterion: $D = -19, -43, -67, -163$. Since the time of Gauss there had been speculation whether there are any other imaginary quadratic fields having unique factorization. In 1934, H. Heilbronn and E. Linfoot proved that there could be at most one more such field. The matter was completely settled in 1966 when it was shown that there existed no other imaginary quadratic field having unique factorization besides the nine we have listed. The first complete proof of this result, by H. M. Stark, makes use of modular functions.

In summary, the problem of unique factorization of integers is now reduced to another, but the new one is far from completely solved. Nevertheless, as we shall see in the sequel, the theory of ideals has far more important consequences than Theorems 9.4 and 9.5.

3. **The problem of ramification.** Let P be a prime ideal in an algebraic number field K. As a generalization

of a fact about Gaussian numbers, we shall show here that P divides exactly one ideal (p) where p is a rational prime.

The proof of Theorem 8.5 shows that there exists a positive rational integer a in P. Let $p_1 \cdots p_r$ be the representation of a as a product of positive rational primes. Then $(a) = (p_1) \cdots (p_r)$, and since $P \mid a$, it follows that $P \mid p$ for some rational prime p.

Now suppose $P \mid p$ and $P \mid q$ where p and q are distinct rational primes. There exist rational integers m and n such that $pm + qn = 1$. Then $P \mid 1$ and $P \supset (1)$, contrary to the fact that P is prime.

It follows that the prime ideals in K can be detected by considering the complete factorization $(p) = P_1 \cdots P_r$ of each ideal (p) into prime ideals P_i of K. An important question which occurs is this: when does (p) have a repeated factor P_i and when are all the P_i distinct? In the former case (p) is said to be *ramified*; otherwise *unramified*. The answer is given by the following theorem of Dedekind. (p) is unramified if and only if $p \nmid d$, where d is the discriminant of K. A complete proof is difficult,* and we shall prove only this part of the theorem: if $p \nmid d$ then (p) is not divisible by the square of a prime ideal.

Let α be an integer in K, and $\alpha_1, \ldots, \alpha_n$ its conjugates for K. We define $S(\alpha)$, the *trace* of α, by

$$S(\alpha) = \alpha_1 + \cdots + \alpha_n .$$

Since $-S(\alpha)$ is the second coefficient of the field polynomial for α, $S(\alpha)$ is a rational integer. Moreover $S(a\alpha) = aS(\alpha)$, for any rational integer a.

It will be useful to express the discriminant d of the field K in terms of the trace. Let $\omega_1, \ldots, \omega_n$ be an integral basis for K and let $\omega_i^{(j)}$ denote the jth conjugate of ω_i.

---

* See, for example, Landau's *Vorlesungen* III.

By the multiplication of determinants

$$d = |\,\omega_i^{(j)}\,|^2 = \begin{vmatrix} \omega_1^{(1)} \cdots \omega_1^{(n)} \\ \cdots\cdots\cdots \\ \omega_n^{(1)} \cdots \omega_n^{(n)} \end{vmatrix} \cdot \begin{vmatrix} \omega_1^{(1)} \cdots \omega_n^{(1)} \\ \cdots\cdots\cdots \\ \omega_1^{(n)} \cdots \omega_n^{(n)} \end{vmatrix}$$

$$= |\sum_j \omega_i^{(j)}\omega_k^{(j)}| = |\,S(\omega_i\omega_k)\,|.$$

Now suppose that (p) has a square factor P^2. We shall prove that $p \mid d$. Let $(p) = P^2Q$. Choose α so that $PQ \mid \alpha$, $P^2Q \nmid \alpha$. Then $\alpha \neq 0$ and $p \nmid \alpha$. Moreover, since $P^2Q \mid P^2Q^2$, $P^2Q^2 \mid (\alpha)^2$, and $(\alpha)^2 = (\alpha^2)$, it follows that $p \mid \alpha^2$. Since $p \geq 2$, $\alpha^2 \mid \alpha^p\beta^p$ for any integer β in K. Hence $p \mid \alpha^p\beta^p$, and $(\alpha\beta)^p/p$ is an integer in K. By the remarks above

$$S((\alpha\beta)^p) = S(p(\alpha\beta)^p/p) = pS((\alpha\beta)^p/p),$$

so that $S((\alpha\beta)^p)$ belongs to (p). Let $\beta_1, \ldots, \beta_n$ be the conjugates of β for K. Then

$$(S(\alpha\beta))^p = (\alpha_1\beta_1 + \alpha_2\beta_2 + \cdots + \alpha_n\beta_n)^p$$

$$= (\alpha_1\beta_1)^p + (\alpha_2\beta_2)^p + \cdots + (\alpha_n\beta_n)^p + p\gamma$$

$$= S((\alpha\beta)^p) + p\gamma$$

where γ is an integer in K. Hence $(S(\alpha\beta))^p$ also belongs to (p) for any integer β in K. Since $S(\alpha\beta)$ is a rational integer, $p \mid S(\alpha\beta)$ in R.

Now let $\omega_1, \ldots, \omega_n$ be an integral basis for K. Then $\alpha = \sum_{k=1}^{n} h_k\omega_k$, where the h_k are rational integers. Since $p \nmid \alpha$ not all the h_k are divisible by p. But

$$S(\alpha\omega_i) = S(\sum_k h_k\omega_k\omega_i) = \sum_k h_kS(\omega_k\omega_i).$$

Since $p \mid S(\alpha\omega_i)$, we can conclude that p divides the last sum. For simplicity let $a_{ki} = S(\omega_k\omega_i)$. We have shown that $|a_{ki}| = d$. Now we shall show that $p \mid d$.

Let A_{ki} be the cofactor of a_{ki} . Then

$$\sum_i A_{ij} \sum_k a_{ki} h_k = \sum_k h_k \sum_i A_{ij} a_{ki} = d h_j .$$

Since p divides each sum $\sum_k a_{ki} h_k$, $p \mid d h_j$ for each j. But not all the h_j are divisible by p. Hence $p \mid d$. We have established the desired result.

THEOREM 9.6. *If* $p \nmid d$, *then* (p) *is unramified.*

COROLLARY 9.7. *Let* $K = R(\zeta)$, *where* ζ *is a primitive* p^{th} *root of unity,* p *an odd rational prime. If* q *is a rational prime and* $q \nmid p$, *then* (q) *is unramified in* K.

This is a consequence of the fact that $q \nmid d$, since $d = (-1)^{(p-1)/2} p^{p-2}$ (Theorem 6.13).

4. Congruences and norms.

Our next aim is to clear the way for a proof of the assertion made in Chapter VII, that an ideal is the totality of integers in K divisible by a fixed integer (not necessarily in K). The reader will find it useful at this stage to review the notion of congruence discussed in §2 of Chapter II.

Let K be a field, A an ideal in K, and α and β integers in K. We define α and β to be *congruent modulo* A (written $\alpha \equiv \beta \pmod{A}$ or $\alpha \equiv \beta(A)$) if $\alpha - \beta$ is in A or, what is the same, if $A \mid (\alpha - \beta)$. The rules for operating with such congruence statements are those set forth in the earlier chapter.

If α is a fixed integer in K, we call the set of all integers congruent to α modulo A a *residue class* modulo A. α is called a *representative* of the class. For example, by Theorem 7.11, every ideal in R must take the form (m), and $0, 1, \ldots, m - 1$ are representatives of the m residue classes modulo (m).

THEOREM 9.8. *If $A \neq (0)$ is an ideal in K, the number of residue classes modulo A is finite.*

Let a be a non-zero rational integer in A. If $\mu \equiv \nu(a)$, then $\mu \equiv \nu(A)$, since $A \supset (a)$. But the number of residue classes in K modulo (a) is finite, as the proof of Lemma 8.4 shows. Since $\mu \not\equiv \nu(A)$ implies $\mu \not\equiv \nu(\bmod\,(a))$, it follows that the number of residue classes modulo A is finite.

Let A be a non-zero ideal in K. The number of residue classes modulo A is called the *norm* of A, written NA or $N(A)$. If A is principal, say $A = (\alpha)$, we write $N((\alpha))$ for the norm, since the notation $N(\alpha)$ can be taken for the norm of the integer α, and the two norms may not be the same in value. Observe that $NA = 1$ if and only if $A = (1)$.

The reader will recall that every non-zero ideal has a basis of integers (Theorem 7.10). We now prove a little more.

LEMMA 9.9. *If $\omega_1, \ldots, \omega_n$ is an integral basis for the algebraic number field K, then each ideal $A \neq (0)$ in it has a basis $\alpha_1, \ldots, \alpha_n$ of the form*

$$\alpha_1 = a_{11}\omega_1,$$
$$\alpha_2 = a_{21}\omega_1 + a_{22}\omega_2,$$
$$\cdots\cdots\cdots\cdots\cdots\cdots\cdots$$
$$\alpha_n = a_{n1}\omega_1 + \cdots + a_{nn}\omega_n,$$

where the a_{ij} are rational integers and all the a_{ii} are positive.

Let a be a non-zero rational integer in A. Then $a\omega_1, \ldots, a\omega_n$ are in A. Let m be fixed, $1 \leq m \leq n$. From all the elements of A which are of the form $a_1\omega_1 + \cdots + a_m\omega_m$, where the a_i are rational integers and $a_m > 0$ (there is at least one such element, since $a\omega_m$ and $-a\omega_m$

are in A), choose an element

$$\alpha_m = a_{m1}\omega_1 + \cdots + a_{mm}\omega_m$$

for which $a_m = a_{mm}$ is least. The α_i, $i = 1, \ldots, n$, so defined have the properties stated in the lemma.

First, the α_i form a basis for K, by Theorem 5.4, since the determinant

$$\begin{vmatrix} a_{11} & 0 & \cdots & 0 \\ a_{21} & a_{22} & 0 & \cdots \\ \cdots\cdots\cdots\cdots\cdots\cdots\cdots \\ a_{n1} & a_{n2} & \cdots & a_{nn} \end{vmatrix} = a_{11}a_{22}\cdots a_{nn}$$

is different from zero. We shall show that the α_i also form a basis for A.

Let α be an integer in A. Since the ω_i form an integral basis for K, we can write

$$\alpha = b_1\omega_1 + \cdots + b_n\omega_n,$$

where the b_i are rational integers. By Theorem 1.1

$$b_n = h_n a_{nn} + r_n, \qquad 0 \le r_n < a_{nn}$$

and therefore

$$\alpha - h_n\alpha_n = \alpha - h_n(a_{n1}\omega_1 + \cdots + a_{nn}\omega_n)$$

$$= b_1'\omega_1 + \cdots + b_{n-1}'\omega_{n-1} + r_n\omega_n$$

is in A. By the definition of a_{nn} we must have $r_n = 0$; then

$$\alpha - h_n\alpha_n = b_1'\omega_1 + \cdots + b_{n-1}'\omega_{n-1}.$$

Now repeat the procedure with b_{n-1}' to obtain

$$\alpha - h_n\alpha_n - h_{n-1}\alpha_{n-1} = b_1''\omega_1 + \cdots + b_{n-2}''\omega_{n-2},$$

and continue until

$$\alpha - h_n\alpha_n - \cdots - h_1\alpha_1 = 0,$$

$$\alpha = h_1\alpha_1 + \cdots + h_n\alpha_n.$$

Hence α can be expressed in terms of the α_i with rational integral coefficients. The representation is unique since the α_i are a basis for K.

This lemma enables us to obtain an explicit formula for the norm of an ideal.

THEOREM 9.10. *If A is a non-zero ideal in K and $\alpha_1, \ldots, \alpha_n$ is a basis for A, then*

$$NA = \left| \frac{\Delta[\alpha_1, \ldots, \alpha_n]}{d} \right|^{1/2},$$

where d is the discriminant of K.

First note that every basis for A has the same discriminant. This follows from the argument used to prove Theorem 6.10. So we can take for the basis of A the one described in the proof of Lemma 9.9. By formula (5.1)

$$\Delta[\alpha_1, \ldots, \alpha_n] = \begin{vmatrix} a_{11} & 0 & \cdots & 0 \\ a_{21} & a_{22} & 0 & \cdots \\ \cdots\cdots\cdots\cdots\cdots\cdots \\ a_{n1} & a_{n2} & \cdots & a_{nn} \end{vmatrix}^2 \Delta[\omega_1, \ldots, \omega_n].$$

By Theorem 6.10, $d = \Delta[\omega_1, \ldots, \omega_n]$, so that

$$\Delta[\alpha_1, \ldots, \alpha_n] = (a_{11}a_{22} \cdots a_{nn})^2 d.$$

The formula of the theorem reduces to $NA = a_{11} \cdots a_{nn}$.

This means we need only show that $a_{11} \cdots a_{nn}$ is the number of distinct residue classes modulo A. For this it suffices to show that

(i) no pair of the $a_{11} \cdots a_{nn}$ numbers

$$r_1\omega_1 + \cdots + r_n\omega_n, \qquad 0 \le r_i < a_{ii},$$

is congruent modulo A;

(ii) every integer in K is congruent to one of these numbers modulo A.

To prove (i) suppose that

$$r_1\omega_1 + \cdots + r_n\omega_n \equiv r_1'\omega_1 + \cdots + r_n'\omega_n(A),$$

where $0 \le r_n < a_{nn}$, $0 \le r_n' < a_{nn}$. We may suppose that $r_n \ge r_n'$. Hence

$$(r_1 - r_1')\omega_1 + \cdots + (r_{n-1} - r_{n-1}')\omega_{n-1}$$
$$+ (r_n - r_n')\omega_n \equiv 0(A).$$

By the definition of a_{nn}, $r_n - r_n' = 0$, $r_n = r_n'$. A similar argument shows that $r_i = r_i'$, $i = 1, \ldots, n-1$.

We prove (ii). Each integer α in the field has the form

$$\alpha = b_1\omega_1 + \cdots + b_n\omega_n$$

for rational integers b_i. Let

$$b_n = h_n a_{nn} + r_n, \qquad 0 \le r_n < a_{nn}.$$

Then

$$\alpha - h_n\alpha_n = b_1'\omega_1 + \cdots + b_{n-1}'\omega_{n-1} + r_n\omega_n.$$

Repeating this procedure with b_{n-1}', b_{n-2}'', $\ldots$, we have

$$\alpha - h_n\alpha_n - \cdots - h_1\alpha_1 = r_1\omega_1 + \cdots + r_n\omega_n$$

where $0 \le r_m < a_{mm}$ for $1 \le m \le n$, so that $\alpha \equiv r_1\omega_1 + \cdots + r_n\omega_n$ modulo A.

COROLLARY 9.11. *If A is principal, $A = (\alpha) \ne 0$, then $NA = |N\alpha|$.*

Clearly $\alpha\omega_1, \ldots, \alpha\omega_n$ is a basis for A, and

$$\Delta[\alpha\omega_1, \ldots, \alpha\omega_n] = (N\alpha)^2\Delta[\omega_1, \ldots, \omega_n] = (N\alpha)^2 d.$$

But by the theorem $\Delta[\alpha\omega_1, \ldots, \alpha\omega_n] = (NA)^2 d$. Hence $(N\alpha)^2 = (NA)^2$. Since $NA > 0$, the corollary follows.

5. **Further properties of norms.** In reading this section the reader will find it instructive to examine each of

our results for the special case $K = R$ and to compare them with the analogous work done in Chapter II.

LEMMA 9.12. *Let A be a non-zero ideal and α and β integers of K. The congruence*

$$\alpha\xi \equiv \beta(A), \qquad ((\alpha), A) = 1,$$

has a solution ξ which is unique modulo A.

Let $\xi_1, \ldots, \xi_{NA}$ be a *complete residue system* modulo A—that is, a set of representatives, one from each residue class. The set of numbers $\alpha\xi_1, \ldots, \alpha\xi_{NA}$ is also a complete residue system. For if $\alpha\xi_1 \equiv \alpha\xi_2$, then $A \mid \alpha(\xi_1 - \xi_2)$. Thus $A \mid (\xi_1 - \xi_2)$ since $((\alpha), A) = 1$, so that $\xi_1 \equiv \xi_2(A)$ and $\xi_1 = \xi_2$. Then among the $\alpha\xi_i$ there is exactly one which is congruent to β modulo A.

THEOREM 9.13. *Let A be a non-zero ideal. The congruence*

$$\alpha\xi \equiv \beta(A)$$

has a solution ξ if and only if $\beta \equiv 0(D)$, where $D = ((\alpha), A)$. If there is a solution it is unique modulo A/D.

If ξ is a solution of the congruence, then $\alpha\xi - \beta = \rho$ is in A and $A \mid \rho$. But then $D \mid \rho$. Also $D \mid \alpha$, so $D \mid \beta$ and β is in D.

Conversely, suppose β is in D. By the definition of D we can find $\alpha\xi$ in (α) and κ in A so that $\alpha\xi + \kappa = \beta$. Then $\alpha\xi \equiv \beta(A)$.

If $\alpha\xi$, $\alpha\xi'$ are both congruent to β, then $\alpha(\xi - \xi') \equiv 0$, $A \mid \alpha(\xi - \xi')$. Let $A = DA_1$, $(\alpha) = DA_2$, where A_1 and A_2 are relatively prime. Then

$$DA_1 \mid DA_2(\xi - \xi'), A_1 \mid A_2(\xi - \xi'), A_1 \mid (\xi - \xi'), \xi \equiv \xi'(A_1),$$

and finally $\xi \equiv \xi'(A/D)$.

THEOREM 9.14. $N(AB) = NA \cdot NB$.

We are assuming that $A, B \neq (0)$. Then according to

Lemma 9.2 it is possible to find γ in A such that $((\gamma)/A, B) = 1$, or $((\gamma), AB) = A$. Let $\alpha_1, \ldots, \alpha_{NA}$ and $\beta_1, \ldots, \beta_{NB}$ be complete residue systems modulo A and B respectively. Then no two of the $NA \cdot NB$ numbers $\alpha_i + \gamma\beta_j$ can be congruent modulo AB. For if $\alpha + \gamma\beta \equiv \alpha' + \gamma\beta'(AB)$, then $\alpha + \gamma\beta \equiv \alpha' + \gamma\beta'(A)$. But γ is in A so that $\alpha \equiv \alpha'(A)$. Since α and α' are elements of a complete residue system, $\alpha = \alpha'$. Hence $\gamma(\beta - \beta') \equiv 0(AB)$. Since $((\gamma), AB) = A$, $(\gamma) = AC$ where $(C, B) = 1$. So $B \mid (\beta - \beta')$. Hence $\beta \equiv \beta'(B)$, $\beta = \beta'$.

To prove the theorem it remains only to show that each integer α in the field K is congruent to one of the numbers $\alpha_i + \gamma\beta_j$ modulo AB. Choose α_i so that $\alpha_i \equiv \alpha(A)$. Now consider the congruence $\gamma\xi \equiv \alpha - \alpha_i(AB)$. By Theorem 9.13 it has a solution since $\alpha - \alpha_i$ is in $A = ((\gamma), AB)$. Moreover ξ can be chosen uniquely modulo $AB/A = B$, so ξ is one of the β_j. Then $\alpha \equiv \alpha_i + \gamma\beta_j(AB)$.

COROLLARY 9.15. *If NA is prime, so is A.*

THEOREM 9.16. *NA is an element of A for $A \neq 0$.*

Let $\alpha_1, \ldots, \alpha_{NA}$ be a complete residue system; $\alpha_1 + 1$, $\ldots, \alpha_{NA} + 1$ is one also, so that

$$\alpha_1 + \cdots + \alpha_{NA} \equiv (\alpha_1 + 1) + \cdots + (\alpha_{NA} + 1)(A),$$

$$0 \equiv NA(A).$$

COROLLARY 9.17. *There are only a finite number of ideals of given norm.*

For NA can belong to only a finite number of ideals (Lemma 8.4).

THEOREM 9.18 (*Fermat's theorem*). *If P is a prime ideal in K and $P \nmid \alpha$, then*

$$\alpha^{NP-1} \equiv 1(P).$$

Let $\alpha_1, \ldots, \alpha_{NP}$ be a complete residue system modulo

P. Then $\alpha\alpha_1, \ldots, \alpha\alpha_{NP}$ is also such a system. One member of each list, say α_{NP} and $\alpha\alpha_{NP}$, is divisible by P. Omitting these and multiplying the other members of each list together, we find that

$$\alpha_1 \cdots \alpha_{NP-1} \equiv \alpha^{NP-1}\alpha_1 \cdots \alpha_{NP-1}(P).$$

Since $P \nmid \alpha_1 \cdots \alpha_{NP-1}$, $1 \equiv \alpha^{NP-1}(P)$.

We have seen in §3 that a prime ideal P divides exactly one rational prime p. We conclude by expressing NP as a power of p.

THEOREM 9.19. *Let P be a prime ideal in a field of degree n over R. Let p be the (unique) rational prime which P divides. Then $NP = p^f$, where $1 \leq f \leq n$.*

By Corollary 9.11, $N((p)) = |Np| = p^n$. Since $P \mid p$, $NP \mid N((p))$, so $NP = p^f$ for some positive integer f not exceeding n.

Problems

1. Let $K = R$. Write (70) and (150) as products of prime ideals. Find $((70), (150))$.
2. Prove that $((1 - 2\sqrt{-5}), (2)) = 1$ in $R(\sqrt{-5})$.
3. Let A, B, and C be ideals. Prove the following assertions:

   (a) $A + B \overset{\text{def}}{=} \{\alpha + \beta; \alpha \in A, \beta \in B\}$ is an ideal;

   (b) $A + B \supset A, A + B \supset B$;

   (c) $C \supset A$ and $C \supset B \Rightarrow C \supset A + B$;

   (d) $A + B = (A, B)$ (the h.c.f. of A and B);

   (e) If $(A, B) = 1$, show that there exist $\alpha \in A$ and $\beta \in B$ such that $\alpha + \beta = 1$. Use this fact to show that if $A \mid BC$ and $(A, B) = 1$, then $A \mid C$.

4. Let A and B be ideals in a field K.

(a) Show that $A \cap B$ is an ideal.

(b) Let $A = P_1^{\alpha_1} \cdots P_r^{\alpha_r}$ and $B = P_1^{\beta_1} \cdots P_r^{\beta_r}$ where the P's are prime ideals and the α's and β's are non-negative rational integers. Show that $A \cap B = P_1^{max(\alpha_1, \beta_1)} \cdots P_r^{max(\alpha_r, \beta_r)}$. $A \cap B$ is called the *least common multiple* of A and B.

(c) Show that $(A \cap B)(A + B) = AB$.

5. Let K be a field and A, B ideals in K each distinct from (1). If $3 \in A \cap B$, can you conclude that $(A, B) \neq 1$?

6. (a) Let $A = (180)$ in R. Use the method of proof of Lemma 9.2 and Theorem 9.3 to find $a \in J$ such that $A = (a, 2^3 \cdot 3^4 \cdot 5^2)$.

   (b) Let $A = (3, 1 + 2\sqrt{-5})$ in $R(\sqrt{-5})$. Using the same method, find an integer α in $R(\sqrt{-5})$ such that $A = (9, \alpha)$.

7. Given an ideal $I = (\alpha, \beta, \gamma)$, it is always possible to express I in terms of two generators. One might conjecture that there always exists a redundant generator, such that I can be expressed in terms of two of the three given generators. Disprove this conjecture by an example in R.

8. Characterize the prime ideals in a field in which factorization of integers is unique.

9. Use the Dedekind-Hasse criterion to show that unique factorization holds in G.

10. Show that the division algorithm fails to hold for any field $R(\sqrt{D})$ with $D \leq -15$.
    Hint. Divide a suitable integer of $R(\sqrt{D})$ by 2.

11. Let p be a rational prime and $K = R(\sqrt{-5})$. Factor (p) in the cases in which it is ramified.

12. Find an algebraic integer α in a quadratic field with $N\alpha = 31$, Trace $\alpha = 17$.

13. Let α and β be non-zero integers of a field K. Prove that $N(\alpha)\, S(\beta/\alpha)$ is a rational integer.

14. Let A be an ideal in a field K. Let $\alpha \equiv \alpha'(A)$ and $\beta \equiv \beta'(A)$. Show that $\alpha\beta \equiv \alpha'\beta'(A)$.

15. Let $K = R(\sqrt{-5})$, $A = (3, 1 + \sqrt{-5})$, $B = (2)$. Show that

 (a) $A = \{3j + (1 + \sqrt{-5})k : j, k \in J\}$.

 (b) 0, 1, 2 are a complete set of residue class representatives modulo A.

 (c) $B = \{2j + 2k\sqrt{-5} : j, k \in J\}$.

 (d) 0, 1, $\sqrt{-5}$, $1 + \sqrt{-5}$ are a complete set of residue class representatives modulo B.

 (e) Show that the discriminant of A is -180 and of B is -320.

16. For each of the ideals (2) and $(2 + i)$ in G

 (a) find a basis;

 (b) draw a diagram showing the ideal in C;

 (c) give a complete set of residue class representatives;

 (d) find the norm;

 (e) find the discriminant.

17. Prove Corollary 9.11 for α a non-zero rational integer by the method used in proving Theorem 9.8.

18. Verify Lemma 9.9 directly for an ideal generated by a single rational integer distinct from zero.

19. Let $\omega_1 = 1 + 2i$, $\omega_2 = 1 + 3i$ be an integral basis for $R(i)$. Find a basis α_1, α_2 for the ideal $(1 + i)$ such that $\alpha_1 = a_{11}\omega_1$, $\alpha_2 = a_{21}\omega_1 + a_{22}\omega_2$ with $a_{11} > 0$, $a_{22} > 0$.

20. Use the technique of Theorem 9.10 to find a complete set of representatives for the residue classes modulo (3) in $R(\sqrt{-5})$.

21. Let $A = (3\sqrt{-5}, 10 + \sqrt{-5})$ in $R(\sqrt{-5})$. Tell whether the congruence $3\xi \equiv 5 \mod A$ has a solution.

22. Let A and B be ideals in a field K, which are each distinct from (0) and (1). Let $\alpha_1, \ldots, \alpha_m$ and $\beta_1, \ldots, \beta_n$ be complete systems of residue class representatives modulo A and B respectively. Can $\{\alpha_i\beta_j : 1 \leq i \leq m, \ 1 \leq j \leq n\}$ serve as a complete system of residue class representatives modulo AB?

23. Using the ideals of problem 15 and the technique of Theorem 9.14, show that γ can be taken as 3 and find a complete set of residue class representatives modulo AB.

24. Let α be an integer of $R(\sqrt{-37})$ and let $I = (2, 1 + \sqrt{-37})$. Show that either α or $\alpha - 1$ is in I.

25. (a) Factor (2) and (5) in $R(\sqrt{-6})$.
    (b) In each case prove that the factors you have found are prime ideals.

26. (a) Let A be an ideal and let m be the smallest positive integer in A. Show that $m \mid NA$.
    (b) If NA is a rational prime, show that $1, 2, \ldots, NA$ constitute a complete set of residue class representatives modulo A.

27. Use Fermat's theorem to solve the congruence

$$(1 + \sqrt{-5})\xi \equiv 3 \mod (7, 1 + 2\sqrt{-5})$$

in $R(\sqrt{-5})$. Explain your reasoning.

28. Does there exist an algebraic number field in which an ideal of norm 12 is a prime?

29. Show that the primes of $R(\sqrt{2})$ are either
    (a) $\pi = a + b\sqrt{2}$ where a and $b \in J$ and $\pm(a^2 - 2b^2)$ is a rational prime or
    (b) of the form $p \neq a^2 - 2b^2$, where p is a rational prime, or an associate of such a rational prime.

CHAPTER X

IDEAL CLASSES AND CLASS NUMBERS

1. Ideal classes. We are nearly ready to justify the assertion that each ideal in $K = R(\theta)$ is the totality of integers in K which are divisible (in the extended sense) by some integer which is not necessarily in K. Our proof will rest on the notion of ideal class.

Two ideals A and B in K are *equivalent*, written $A \sim B$, if there are two non-zero integers α and β in K such that

$$(\alpha)A = (\beta)B.$$

The simplest properties of this equivalence relation are the following:

   (i) $A \sim A$;
   (ii) $A \sim B$ if and only if $B \sim A$;
   (iii) if $A \sim B$ and $B \sim C$, then $A \sim C$;
   (iv) all non-zero principal ideals are equivalent.

The totality of ideals in K equivalent to a fixed ideal $A \neq (0)$ is said to constitute a *class*. The number of classes (which we shall soon show to be finite) is called the *class-number* h of K. If the class-number is 1 then all non-zero ideals are equivalent to (1) and so are all principal. From Theorem 9.4 it follows that a field has unique factorization of integers into prime integers if and only if its class-number is 1.

LEMMA 10.1. *Let A, B, C be ideals in K and $A \neq (0)$. Then $AB \sim AC$ if and only if $B \sim C$.*

The "if" implication is immediate from the definition of

equivalence. Conversely, suppose $AB \sim AC$. Choose D to be an ideal for which AD is principal. Then $AD \sim (1)$, and $(1) B \sim ADB \sim ADC \sim (1)C$.

2. Class numbers.

The finiteness of the class number of any field will be established using

THEOREM 10.2. *Let K be a field. There exists a constant C, depending only on K, such that any non-zero ideal A in K contains an element $\alpha \neq 0$ with $| N(\alpha) | \leq CN(A)$.*

Let $\omega_1, \ldots, \omega_n$ be an integral basis for K and take t to be the greatest rational integer not exceeding $(NA)^{1/n}$. Consider the set of numbers $\beta = b_1\omega_1 + \cdots + b_n\omega_n$ with b_i rational integers satisfying $0 \leq b_i \leq t$ for $1 \leq i \leq n$. Since $(t + 1)^n > NA$, there exist at least two of the numbers β, β' which are congruent modulo A. Let $\alpha = \beta - \beta' = a_1\omega_1 + \cdots + a_n\omega_n$. Then $| N\alpha | = \prod_{j=1}^{n} | \sum_{i=1}^{n} a_i\omega_i^{(j)} | \leq t^n \prod_{j=1}^{n} \sum_{i=1}^{n} | \omega_i^{(j)} | \leq NA \cdot C$.

Another argument, based on the Minkowski lemma on linear forms, would allow us to take $C = \sqrt{|d|}$, where d is the discriminant of K. Moreover, for $K \neq R$, the inequality can be shown to be strict.

THEOREM 10.3. *The class number h of a field is finite.*

It suffices to show that each class of ideals contains an ideal B of norm at most C, for by Corollary 9.17 there are at most a finite number of ideals having a given norm.

Let a class be given and let D be any ideal in it. Choose A so that AD is principal. Then $AD \sim (1)$. By the preceding theorem there exists an $\alpha \neq 0$ in A such that $| N\alpha | \leq CNA$. Since $A | \alpha$, $(\alpha) = AB$ for some ideal B, and $N((\alpha)) = | N\alpha | = NA \cdot NB$. Consequently, $NB \leq C$. Finally, since $AB \sim (1) \sim AD$, it follows that B lies in the given class.

We pause to compute the class number of a specific

field. We shall show that the class number of $R(\sqrt{-5})$ is 2, i.e., in this case there are just two classes, the principal and the non-principal ideals.

We begin by giving a numerical estimate for the constant C in Theorem 10.2 for $K = R(\sqrt{-5})$. The numbers 1 and $\sqrt{-5}$ form a basis for this field. The *proof* of Theorem 10.2 shows that there exists an $\alpha \neq 0$ with

$$|N\alpha| = |a_1 + \sqrt{-5}a_2||a_1 - \sqrt{-5}a_2|$$

$$= a_1^2 + 5a_2^2 \leq 6NA.$$

The proof of the preceding theorem showed that each class contains an ideal of norm at most 6. Thus we shall consider ideals having norms $1, \ldots, 6$.

There is just one ideal of norm 1, namely (1).

Any ideal of norm 2, 3, or 5 must be prime, and by Theorem 9.19 it must occur in the factorization $(p) = P_1^{e_1} \cdots P_r^{e_r}$. Here $p = 2$, 3, or 5, the P_i are distinct prime ideals, and $e_i \geq 1$ for $1 \leq i \leq r$. Taking norms, we obtain $p^2 = (NP_1)^{e_1} \cdots (NP_r)^{e_r}$. Clearly there are just three possible cases: $(p) = P$, $(p) = P^2$, and $(p) = PP'$, $P \neq P'$. One verifies that $(2) = P_2^2$, where $P_2 = (2, 1 + \sqrt{-5})$; $(3) = P_3 P_3'$, where $P_3 = (3, 1 + 2\sqrt{-5}) \neq (3, 1 - 2\sqrt{-5}) = P_3'$; and $(5) = (\sqrt{-5})^2$. By unique prime factorization, P_2, P_3, P_3', and $(\sqrt{-5})$ are the only ideals having norms 2, 3, or 5.

An ideal of norm 4 has to be the product of prime ideals dividing (2) by Theorem 9.19. Thus $P_2^2 = (2)$ is the only ideal of norm 4. Similarly, an ideal of norm 6 is the product of prime ideals dividing (2) and (3). Thus $P_2 P_3 = (1 - \sqrt{-5})$ and $P_2 P_3' = (1 + \sqrt{-5})$ are the only ideals of norm 6.

We have now found all ideals of norm at most 6. Clearly, $(1) \sim (\sqrt{-5}) \sim (2) \sim (1 - \sqrt{-5}) \sim$

$(1 + \sqrt{-5})$. It remains to verify that $P_2 \sim P_3 \sim P'_3 \not\sim (1)$. However, we have noted already that $P_2^2 \sim P_2 P_3 \sim P_2 P'_3 \sim (1)$ and hence $P_2 \sim P_3 \sim P'_3$. Also, $P_3 \not\sim (1)$ since P_3 is not a principal ideal (§7.4).

COROLLARY 10.4. *If A is an ideal in K, and h is the class-number of K, then A^h is principal.*

If $A = (0)$, $A^h = (0)$ and the result is clear. Suppose that $A \neq (0)$. Choose a set of ideals $A_1, \ldots, A_h$, one from each class in K. Then $AA_1, \ldots, AA_h$ fall into distinct classes, for if $AA_i \sim AA_j$ then $A_i \sim A_j$. Hence

$$A_1 \cdots A_h \sim AA_1 \cdot AA_2 \cdots AA_h = A^h A_1 \cdots A_h ,$$

so $A^h \sim (1)$ and A^h is principal.

COROLLARY 10.5. *Let q be a positive rational integer which is relatively prime to h. If $A^q \sim B^q$, then $A \sim B$.*

Since h and q are relatively prime, there exist rational integers r and s such that $qr - hs = 1$. Also, $q(r + nh) - h(s + nq) = 1$ holds for any n. Thus we may assume that r and s are both positive.

Now $A^{qr} \sim B^{qr}$ and thus $A^{hs}A \sim B^{hs}B$. But A^h and B^h are principal, and hence so are A^{hs} and B^{hs}. It follows that $A \sim B$.

We shall now prove that any non-zero ideal in K is the totality of integers α in K which are divisible by a fixed integer κ, not necessarily in K. The integer κ need not be unique, even to within units. For example, let $A = (2)$ in R, the rational field. Then A consists of all the even rational integers—that is, the totality of integers in R divisible by 2. But A is also the totality of integers in R divisible by $\sqrt{2}$ (in the extended sense of division). For $n/\sqrt{2}$ is an algebraic integer if n is even, but not if n is odd. But there is uniqueness in this sense: among all the κ which have the desired property there is one which is divisible by all the others; *this* one is unique to within units. In the special case just considered, the integer 2 is the one which contains all other κ as factors. Of course -2 serves equally well.

THEOREM 10.6. *For each non-zero ideal A in K there is an integer κ, not necessarily in K, such that*

(i) *A is the totality of integers δ in K for which δ/κ is integral;*

(ii) *if κ' is an integer which divides every element of A, then κ' divides κ.*
κ is unique to within units.

Let $A = (\alpha, \beta)$. Then $(\alpha, \beta)^h = (\omega)$ is principal, by Corollary 10.4. $\kappa = \omega^{1/h}$ is an integer since it satisfies the equation $x^h - \omega = 0$. Consider the extension $E = K(\kappa)$ of K. E contains K and hence all the elements of A. Now in K

$$A^h = (\alpha, \beta)^h = (\omega).$$

By Theorem 8.1 these ideals are equal when considered as ideals in any finite extension of K. Then $(\alpha, \beta)^h = (\omega) = (\kappa)^h$ in E. In view of the unique factorization theorem for ideals in E, $(\alpha, \beta) = (\kappa)$, still in E. Hence every element of A is divisible by κ. Moreover

(10.1) $$\kappa = \lambda\alpha + \nu\beta,$$

where λ and ν are in E.

Conversely we must show that any element γ in K which is divisible by κ is in A. Since γ and κ are both in E, and $\kappa \mid \gamma$ it follows that γ is in $(\kappa) = (\alpha, \beta)$, *where (α, β) is considered as an ideal in E.* We wish to show that γ is in (α, β) when (α, β) is considered as an ideal in K—this is not yet clear. But $\gamma = \lambda\kappa$, where λ is an integer in E. Let κ be of degree k over K. Then $E = K(\kappa)$,

$(E/K) = k$. Let $\kappa_1, \kappa_2, \ldots, \kappa_k$ denote the conjugates of κ and $\lambda_1, \ldots, \lambda_k$ the conjugates of λ for E. γ is in K so that all its conjugates are the same. Hence

$$\gamma = \lambda_i \kappa_i, \, i = 1, \ldots, k; \qquad \gamma^k = (\lambda_1 \cdots \lambda_k)(\kappa_1 \cdots \kappa_k).$$

The product $\xi = \lambda_1 \cdots \lambda_k$ is symmetric in the λ_i, so it is an integer in K. Since κ satisfies $x^h - \omega = 0$, so does each of $\kappa_1, \ldots, \kappa_k$. Then

$$\kappa_i^h = \omega, \qquad (\kappa_1 \cdots \kappa_k)^h = \omega^k, \qquad \gamma^{hk} = \xi^h \omega^k;$$

hence, as ideals in K,

$$(\gamma)^{hk} = (\xi)^h(\omega)^k = (\xi)^h A^{hk},$$

$$(\gamma)^k = (\xi)A^k, \qquad A^k \mid (\gamma)^k$$

by the fundamental theorem of ideal theory. By another application of this theorem it follows that $A \mid \gamma$, so that γ is in A. This proves part (i) of the theorem, and (ii) follows from (10.1).

To prove the uniqueness of κ suppose that κ_1 and κ_2 both have properties (i) and (ii). Then $\kappa_1 \mid \kappa_2$, $\kappa_2 \mid \kappa_1$ so that $\kappa_2 = \sigma \kappa_1$, where σ is a unit.

Problems

1. Prove that multiplication of ideal classes is well defined. That is, show that if $A \sim B$ and $C \sim D$ then $AC \sim BD$.
2. Show that the ideal classes of a given field form an abelian group under multiplication.
3. Show that the class number of $R(\sqrt{3})$ is one.
4. Which of the following ideals in $R(\sqrt{-6})$ are equivalent? $(2, \sqrt{-6})$, $(5 - 2\sqrt{-6}, 2 + 2\sqrt{-6})$, $(2 + \sqrt{-6})$, $(5, 1 + 2\sqrt{-6})$.
5. Give an upper estimate for the class number of $R(\sqrt{-6})$.

6. Let A be an ideal and let r be the minimal positive integer for which A^r is principal. Show that r divides the class number.

7. Let $(\alpha, \beta) = A$, an ideal in a field K, and let $A^2 \sim (1)$ in K. Suppose that $E \supset K$ and let $(\alpha, \beta) = B$, an ideal in E. Prove that $B^2 \sim (1)$ in E.

8. Find an integer κ such that the ideal $(3, 1 + 2\sqrt{-5})$ is the totality of integers in $R(\sqrt{-5})$ which are divisible by κ in the extended sense.

CHAPTER XI

THE FERMAT CONJECTURE

1. Pythagorean triples. The equation $x^2 + y^2 = z^2$ has solutions in rational integers with $x \neq 0$ and $y \neq 0$, e.g., $3^2 + 4^2 = 5^2$. A solution in rational integers is called *primitive* if x, y, and z are relatively prime. Clearly, a solution of $x^2 + y^2 = z^2$ is primitive if and only if any two of x, y, z are relatively prime. If (x, y, z) is primitive, one of x, y must be odd. Moreover, both could not be odd, for otherwise $x^2 + y^2 \equiv 2 \pmod 4$ and the congruence $z^2 \equiv 2 \pmod 4$ has no solution. The following theorem characterizes the primitive solutions of this equation with $x > 0$, $y > 0, z > 0$.

THEOREM 11.1. *All positive primitive solutions of $x^2 + y^2 = z^2$ with y even are given by $x = r^2 - s^2$, $y = 2rs$, $z = r^2 + s^2$, where r and s range over relatively prime rational integers of opposite parity satisfying $r > s > 0$.*

Let x, y, z be a positive primitive solution of $x^2 + y^2 = z^2$. Then $z^2 = (x + iy)(x - iy)$ holds in G. We shall show that $x + iy$ and $x - iy$ are each squares and derive the desired formulas for x and y.

We must first show that $x + iy$ and $x - iy$ are relatively prime. Since x and y are relatively prime rational integers, there exist integers u and v such that $ux + vy = 1$. Thus x and y are also relatively prime in G. Let π be a prime in G, $\pi \mid x + iy$, $\pi \mid x - iy$. Then $\pi \mid 2x$, $\pi \mid 2y$ and hence $\pi \mid 2$. Consequently $\pi = 1 + i$ or an associate. But $1 + i \nmid x + iy$, since $N(1 + i) = 2$ and $N(x + iy) = x^2 + y^2$ is odd. Thus $(x + iy, x - iy) = 1$.

It follows by unique factorization that each of $x + iy$ and $x - iy$ is a square, e.g., $x + iy = \epsilon(c + id)^2$ for some $c + id \in G$ and $\epsilon = \pm 1$ or $\pm i$. Thus

$$x + iy = \epsilon(c^2 - d^2 + 2icd).$$

Since x is odd and y is even, $y = \pm 2cd$ and so $\epsilon = \pm 1$. It follows also that c and d are of opposite parity.

It remains to consider the two cases $\epsilon = +1$ and $\epsilon = -1$. If $\epsilon = +1$, then $c^2 > d^2$ since $x > 0$, and $cd > 0$ since $y > 0$. In this case, $r = |c|$, $s = |d|$. If $\epsilon = -1$, then $c^2 < d^2$ and $cd < 0$ and we take $r = |d|$, $s = |c|$. In both cases the r and s satisfy the assertions of the theorem.

Conversely, if x, y, and z are expressed in terms of r and s satisfying the stated conditions, then $x^2 + y^2 = z^2$. Also, x, y, z are relatively prime, for if p is a rational prime and $p \mid x$ and $p \mid z$, then $p \mid 2r$ and $p \mid 2s$. Hence $p \mid 2$ and so $p = 2$. But $2 \nmid r^2 - s^2$ since r and s are of opposite parity.

2. The Fermat Conjecture.

The reader is probably acquainted with the following famous problem: for what positive integral values of n does the equation

$$(11.1) \qquad x^n + y^n = z^n$$

have a solution in non-zero rational integers x, y, z? We have seen that there are such non-trivial solutions for $n = 2$. It was asserted by Fermat in 1637 that if $n > 2$, there is no non-trivial solution (Fermat's "Last Theorem"), but a proof has never been found and the assertion at present has only the status of a conjecture. A large part of the theory of algebraic numbers originated in an effort to prove it.

Fermat did show the impossibility of solving (11.1) non-trivially for $n = 4$. We shall establish this fact using

Theorem 11.1. Indeed we show somewhat more in

THEOREM 11.2. *The equation* $x^4 + y^4 = z^2$ *has no solution in positive rational integers* x, y, z.

If the assertion of the theorem were false, there would exist a solution having a minimal value of z. We shall assume the existence of such a solution and show that it leads to another solution having a smaller value of z. This argument is known as the "method of descent" and is due to Fermat.

We must have $(x, y) = 1$, for otherwise we could divide through the equation by $(x, y)^4$. If both of x, y are odd, then $z^2 = x^4 + y^4 \equiv 2 \pmod 4$. But the congruence $z^2 \equiv 2 \pmod 4$ has no solution, and hence one of x, y is odd and the other even.

Supposing y to be even and x odd, we apply Theorem 11.1 to $(x^2)^2 + (y^2)^2 = z^2$. We have

$$x^2 = r^2 - s^2, \qquad y^2 = 2rs, \qquad z = r^2 + s^2$$

with $r > s > 0$, $(r, s) = 1$, and $r + s$ odd. If r were even and s odd, then $x^2 \equiv -1 \pmod 4$, an unsolvable congruence. Thus we can express $s = 2t$.

Now $y^2 = 2rs = 4rt$, and $(r, t) = 1$. Thus $r = \rho^2$ and $t = \tau^2$ for some relatively prime positive integers ρ and τ. Also,

$$x^2 = r^2 - s^2 = \rho^4 - 4\tau^4, \quad x^2 + 4\tau^4 = \rho^4.$$

Once again we apply Theorem 11.1, this time to $x^2 + (2\tau^2)^2 = (\rho^2)^2$. Note that $(2\tau^2, \rho^2) = 1$ and thus $(x, 2\tau^2, \rho^2)$ is a primitive solution. Thus we have $2\tau^2 = 2lm$, $\rho^2 = l^2 + m^2$ with $(l, m) = 1$. It follows that $l = \lambda^2$, $m = \mu^2$ for some positive integers λ and μ and hence $\rho^2 = \lambda^4 + \mu^4$. Also, $\rho^2 = r < y^2 < z^2$. This contradicts the minimality of z and proves the theorem.

Before discussing the conjecture further we shall simplify its statement somewhat. Since there is no positive solution for $n = 4$ there can be no such solution when $n = 4m$, for we can write (11.1) as $(x^m)^4 + (y^m)^4 = (z^m)^4$. Every integer $n \neq 4m$, $n > 2$, can be written in the form $n = pr$, where p is an odd prime; hence it is enough to show that (11.1) has no non-trivial solution when n is an odd prime. For we can write $(x^r)^p + (y^r)^p = (z^r)^p$. Finally, we let $n = p$ and replace z by $-z$; since p is odd (11.1) becomes

$$(11.2) \qquad x^p + y^p + z^p = 0.$$

Fermat's conjecture then amounts to this: for no odd prime p does (11.2) have a non-trivial solution in rational integers.

It is convenient to classify the primes p as follows. Let h be the class-number of $R(\zeta)$, where ζ is a primitive p^{th} root of unity. If $p \nmid h$, p is *regular*; otherwise p is *irregular*. For example, let $K = R(\zeta)$, where ζ is a non-real cube root of unity. Then $K = R(\sqrt{-3})$, a field which is known to be Euclidean (cf. §9.2). Thus all ideals of K are principal and the class number is 1. Since $3 \nmid 1$, 3 is a regular prime.

Kummer proved that if p is regular, then (11.2) has no non-trivial solution in rational integers. Unfortunately there are an infinite number of irregular primes (of which $p = 37$ is the smallest), and at present the Fermat problem has been solved only for certain of these primes. We shall illustrate the connection of Fermat's conjecture with algebraic number theory by proving a weakened version of Kummer's theorem. The reader wishing further information on this subject should consult Borevich-Shafarevich, Landau III, or Vandiver's expository paper "Fermat's Last Theorem", American Math. Monthly, vol. 53 (1946), pp. 555–578.

3. **Units in cyclotomic fields.** Let ζ denote a primitive p^{th} root of unity, $p \neq 2$. Let $K = R(\zeta)$, a cyclotomic field of degree $p - 1$ over R. We have set out some properties of K in Chapters V and VI. Here we shall prove a series of lemmas leading to a characterization of units in cyclotomic fields. To avoid a real danger of confusion, ideals will be written in *square brackets* rather than parentheses. As earlier, λ will denote $1 - \zeta$. L denotes the ideal $[\lambda]$.

LEMMA 11.3. *$1 - \zeta$ and $1 - \zeta^j$ are associates for $1 \leq j \leq p - 1$.*

$1 - \zeta^j = (1 - \zeta)(1 + \zeta + \cdots + \zeta^{j-1})$ and hence $1 - \zeta \mid 1 - \zeta^j$. Conversely, for $1 \leq j \leq p - 1$ there exists a rational integer t, $1 \leq t \leq p - 1$ such that $jt \equiv 1$ (mod p). Then $1 - \zeta = 1 - \zeta^{tj}$ and as above,

$$1 - \zeta^j \mid 1 - \zeta^{tj}.$$

LEMMA 11.4. *$L^{p-1} = [p]$ and $NL = p$.*

As we proved in Chapter V,

$$p = (1 - \zeta)(1 - \zeta^2) \cdots (1 - \zeta^{p-1}).$$

The preceding lemma implies that $[1 - \zeta^j] = [1 - \zeta]$ for $1 \leq j \leq p - 1$. Thus $[p] = [1 - \zeta]^{p-1} = L^{p-1}$.

Since $(K/R) = p - 1$, $N[p] = |Np| = p^{p-1}$. Then $(NL)^{p-1} = p^{p-1}$, $NL = p$. By Corollary 9.15 L is a prime ideal.

LEMMA 11.5. *The number i is not in K, nor is the number $e^{2\pi i/q}$ if q is a prime different from p and greater than 2.*

Suppose i is in K. Since i is a unit $[1 + i] = [1 - i]$. Then

$$[2] = [1 + i][1 - i] = [1 + i]^2.$$

Since $2 \neq p$ this contradicts Corollary 9.7.

If $e^{2\pi i/q}$ is in K then, by the same argument used to prove Lemma 11.4,

$$[q] = [1 - e^{2\pi i/q}]^{q-1}.$$

Since $q > 2$, $[q]$ is ramified, again in contradiction to Corollary 9.7.

By the fundamental theorem of algebra, the polynomial equation $f(x) = x^m - 1 = 0$ has m roots. An mth root of unity is a number α satisfying $f(\alpha) = \alpha^m - 1 = 0$. The collection of numbers $1, e^{2\pi i/m}, \ldots, e^{2(m-1)\pi i/m}$ gives m roots of f, and thus each mth root of unity is of this form.

LEMMA 11.6. *The only roots of unity in K are $\pm\zeta^s$, $0 < s \le p$.*

Suppose $\alpha = e^{2\pi it/m}$ is in K. We can assume that $m > 0$, $(m, t) = 1$. Let $\zeta = e^{2\pi ij/p}$ for some $j = 1, \ldots, p - 1$. The lemma asserts that there exist rational integers s and k such that

$$\frac{2\pi it}{m} = \frac{2\pi isj}{p} + k\pi i, \quad \text{or} \quad \frac{2pt}{m} = 2sj + kp.$$

If $m \mid 2pt$, then we can find such s and k, since $2sj$ runs through a complete set of residue class representatives mod p for $s = 1, 2, \ldots, p$. Since $(m, t) = 1$, it suffices to show that $m \mid 2p$. If $m \nmid 2p$, then one of the following must be true:

$$4 \mid m, \quad q \mid m, \quad \text{or} \quad p^2 \mid m,$$

where q is an odd prime different from p.

Since $(m, t) = 1$ we can find r so that $tr \equiv 1 \pmod{m}$, $tr = 1 + km$. Then

$$\alpha^r = e^{2\pi itr/m} = e^{2\pi i(k+1/m)} = e^{2\pi i/m}$$

is in K.

If $4 \mid m$ then $e^{2\pi i/4} = i$ is in K, contradicting Lemma 11.5. If $q \mid m$ then $e^{2\pi i/q}$ is in K, also contradicting that lemma.

If $p^2 \mid m$ then $\tau = e^{2\pi i/p^2}$ is in K. We show that this is impossible. τ satisfies the equation $x^{p^2} - 1 = 0$, but not $x^p - 1 = 0$. Hence τ is a root of

$$\frac{x^{p^2} - 1}{x^p - 1} = x^{p(p-1)} + x^{p(p-2)} + \cdots + 1.$$

By Theorem 3.9, τ is of degree $p(p-1) > p-1$ over R, hence (Corollary 5.8) it cannot belong to K which is of degree $p-1$ over R.

LEMMA 11.7. *For each integer α in K there is a rational integer a such that*

$$\alpha^p \equiv a \pmod{L^p}.$$

Since $NL = p$, there are p incongruent residue classes modulo L. The numbers $0, 1, \ldots, p-1$ form a complete system of residue class representatives modulo L, since they are mutually incongruent modulo L. Indeed, say that $1 \le a < b \le p-1$ and $b - a \in L$. Also, $p \in L$ by Theorem 9.16. Now $(b - a, p) = 1$ and hence $1 \in L$, which is impossible since L is prime.

Hence for a suitable rational integer b, $\alpha \equiv b \ (L)$. Now

$$\alpha^p - b^p = \prod_{m=0}^{p-1} (\alpha - \zeta^m b).$$

Since $\lambda = 1 - \zeta$, $\zeta \equiv 1(L)$ and each of the factors

$$\alpha - \zeta^m b \equiv \alpha - b \equiv 0 \ (L).$$

Hence $\alpha^p - b^p \equiv 0 \ (L^p)$.

LEMMA 11.8. *If all the coefficients of a monic polynomial are rational integers and all the roots are of absolute value 1 then these roots are roots of unity.*

Let the roots be $\omega_1, \ldots, \omega_k$. For each positive integer l let

$$p_l(x) = (x - \omega_1^l)(x - \omega_2^l) \cdots (x - \omega_k^l),$$

a monic polynomial, also expressible as

$$p_l(x) = x^k + a_{l,k-1}x^{k-1} + \cdots + a_{l,0}.$$

For each l, $(-1)^j a_{l,k-j}$ denotes the jth elementary symmetric function in $\omega_1^l, \ldots, \omega_k^l$. Each $a_{l,j}$ is a rational integer by Theorem 3.10. Also, since the roots are of absolute value 1,

$$|a_{l,j}| \leq \binom{k}{j}$$

independently of l. Thus there can be at most a finite number of different polynomials $p_l(x)$. It follows that there exist integers $l < m$ such that $p_l(x) = p_m(x)$ and moreover the roots agree pairwise: $\omega_1^l = \omega_1^m, \ldots, \omega_k^l = \omega_k^m$. Then $\omega_i^{m-l} = 1$ and ω_i is a root of unity for $1 \leq i \leq k$.

LEMMA 11.9. *Let ϵ be a unit in $R(\zeta)$. Then $\epsilon = \zeta^g r$, where g is a positive rational integer and r is a real number.*

Since $1, \zeta, \ldots, \zeta^{p-2}$ is an integral basis, $\epsilon = r(\zeta)$, where $r(\zeta)$ is a polynomial in ζ with rational integral coefficients. For $s = 1, \ldots, p-1$ the number $\epsilon_s = r(\zeta^s)$ is conjugate to ϵ. Since $N\epsilon = \epsilon_1 \cdots \epsilon_{p-1} = \pm 1$, $\epsilon_s \mid 1$, so that each ϵ_s is a unit. Moreover,

$$\epsilon_{p-s} = r(\zeta^{p-s}) = r(\zeta^{-s}) = \overline{r(\zeta^s)},$$

where the bar denotes the complex-conjugate. Hence $\epsilon_{p-s} = \bar{\epsilon}_s$, $\epsilon_s \epsilon_{p-s} = |\epsilon_s|^2 > 0$. There are $p-1$ of the ϵ_s; multiplying them in pairs we get $N\epsilon = \prod \epsilon_s \epsilon_{p-s} > 0$, so that $N\epsilon = 1$.

The numbers $\epsilon_s/\epsilon_{p-s}$, $s = 1, \ldots, p-1$, are units of absolute value 1. By the usual argument on symmetric functions the polynomial

$$\prod_{s=1}^{p-1}\left(x - \frac{\epsilon_s}{\epsilon_{p-s}}\right) = \prod_{s=1}^{p-1}(\epsilon_{p-s}x - \epsilon_s)$$

has rational integral coefficients. We conclude from Lemma 11.8 that $\epsilon_s/\epsilon_{p-s}$ is a root of unity. In particular if we let $s = 1$, we find that ϵ/ϵ_{p-1} is a root of unity. By Lemma 11.6, $\epsilon/\epsilon_{p-1} = \pm\zeta^t = \pm\zeta^{t+p}$. Since p is odd one of t or $t + p$ is even, so that $\epsilon/\epsilon_{p-1} = \pm\zeta^{2g}$, where g is a positive rational integer.

Modulo L the numbers $0, 1, \ldots, p - 1$ form a complete residue system. Hence for some one of them, v,

(11.3) $\zeta^{-g}\epsilon \equiv v \ (L)$.

But $L = [\lambda]$, so that $\mu = (\zeta^{-g}\epsilon - v)/\lambda$ is an integer in K. Its complex-conjugate $\bar{\mu}$ is also an integer in K for both satisfy the same minimal polynomial. Then

$$\bar{\mu} = \frac{\overline{\zeta^{-g}}\bar{\epsilon} - v}{\bar{\lambda}} = \frac{\zeta^{g}\epsilon_{p-1} - v}{\bar{\lambda}}$$

is an integer. But $\bar{\lambda} = 1 - \zeta^{p-1}$ is an associate of λ by Lemma 11.3. Hence $(\zeta^{g}\epsilon_{p-1} - v)/\lambda$ is also an integer. Then

$$\zeta^{g}\epsilon_{p-1} \equiv v \equiv \zeta^{-g}\epsilon \ (L)$$

by (11.3). This shows that $\epsilon/\epsilon_{p-1} \equiv \zeta^{2g} \ (L)$.

We can now decide for which choice of sign our previous conclusion $\epsilon/\epsilon_{p-1} = \pm\zeta^{2g}$ is correct. If the $-$ sign holds then $-\zeta^{2g} \equiv \zeta^{2g} \ (L)$, so that $L \mid 2\zeta^{2g}$, $NL \mid 2^{p-1}$, which contradicts Lemma 11.4. This means that $\epsilon = \zeta^{2g}\epsilon_{p-1}$, $\epsilon\zeta^{-g} = \epsilon_{p-1}\zeta^{g}$. Since the right- and left-hand sides of this equation are complex-conjugates and equal, they must be real. The lemma is established.

4. **Kummer's theorem.** We are now in a position to prove the following simplified form of Kummer's theorem:

THEOREM 11.10. *If p is a regular odd prime, then*

$$(11.4) \qquad x^p + y^p + z^p = 0$$

has no solution in rational integers for which $p \nmid xyz$.

The theorem is in fact true with the weaker condition $xyz \neq 0$ in place of $p \nmid xyz$, but the proof is more intricate.

We shall assume that the equation has a solution for which p does not divide any of h, x, y, z and arrive at a contradiction. If x and y have a common factor it is shared by z, and we can remove it by division. This justifies the assumption that x, y, and z have no common factor. From (11.4) we obtain

$$(11.5) \qquad \prod_{m=0}^{p-1} (x + \zeta^m y) = -z^p,$$

and then

$$(11.6) \qquad \prod_{m=0}^{p-1} [x + \zeta^m y] = [z]^p.$$

(Note the change in notation, signifying the passage from numbers to ideals.)

Each two of the ideals on the left of (11.6) are relatively prime. To prove this suppose P is a prime factor of both $[x + \zeta^k y]$ and $[x + \zeta^l y]$, $0 \leq k < l \leq p - 1$. Then P contains both $x + \zeta^k y$ and $x + \zeta^l y$, and hence their difference $y\zeta^k(1 - \zeta^{l-k})$. Since $1 - \zeta^{l-k}$ is an associate of $1 - \zeta = \lambda$ and ζ^k is a unit, P contains the number $y\lambda$. Hence P contains either y or λ, $P \mid y$ or $P \mid \lambda$. In addition, by (11.6) $P \mid z$, so P contains z and therefore P contains $-z^p = x^p + y^p$. There are now two possibilities. (i) If P contains y it contains $x^p = -z^p - y^p$, so $P \mid x$, $P \mid y$, contrary to the fact that x and y are relatively prime. (ii) If $P \mid \lambda$ then $P \mid L$. But L is prime, so that $P = L$. Then L contains z, $L \mid z$, $NL \mid Nz$, $p \mid z^{p-1}$, $p \mid z$, contrary to the hypothesis that $p \nmid z$. This proves the assertion made at the beginning of the paragraph.

We return to (11.6). Since the factors on the left-hand side are relatively prime, it follows from the fundamental theorem of ideal theory that each of them must be the pth power of an ideal. In particular $[x + \zeta y] = A^p$. Then A^p is principal, $A^p \sim (1)$ and, by Corollary 10.5, $A \sim (1)$. Hence A is itself a principal ideal (δ), and $[x + \zeta y] = [\delta]^p = [\delta^p]$. This shows that $x + \zeta y = \epsilon \delta^p$, where ϵ is a unit.

This step we have just taken is the decisive one. But could we not have drawn the same conclusion directly from (11.5) without the excursion into ideal theory? The answer is that without ideal theory we could have made the direct step only in the case when the field $R(\zeta)$ has class number 1—that is, when factorization of *integers* into primes is unique. Unfortunately there are cyclotomic fields of class number greater than 1.

Since $x + \zeta y = \epsilon \delta^p$ we can invoke Lemma 11.9 to conclude that $x + \zeta y = \zeta^g r \delta^p$, where r is a real number. According to Lemma 11.7, $\delta^p \equiv a (L^p)$ for some rational integer a. Hence $x + \zeta y \equiv \zeta^g r a\ (L^p)$. But (Lemma 11.4) $[p] \mid L^p$, and therefore $x + \zeta y \equiv \zeta^g r a\ (\text{mod } [p])$. Since ζ^{-g} is a unit $\zeta^{-g}(x + \zeta y) \equiv r a$. Also $r a$ is real, so that by taking complex-conjugates we find that $\zeta^g (x + \zeta^{-1} y) \equiv r a$. Combining the last two congruences yields

$$(11.7) \quad x\zeta^{-g} + y\zeta^{1-g} - x\zeta^g - y\zeta^{g-1} \equiv 0\ (\text{mod } [p]).$$

Observe next that $g \not\equiv 0\ (\text{mod } p)$. For otherwise $\zeta^g = 1$ and (11.7) becomes $y(\zeta - \zeta^{-1}) \equiv 0$, $y(1 + \zeta)(1 - \zeta) \equiv 0$; then $p \mid y\lambda$ since $1 + \zeta = (1 - \zeta^2)/(1 - \zeta)$ is a unit by

Lemma 11.3. Since $[p] = [\lambda]^{p-1}$, $p > 2$, $\lambda^2 \mid y\lambda$, $\lambda \mid y$, $N\lambda \mid Ny$, $p \mid y^{p-1}$, contrary to the hypothesis $p \nmid y$. Similarly $g \not\equiv 1 \pmod{p}$. For otherwise (11.7) becomes $x(\zeta^{-1} - \zeta) \equiv 0$, and a similar argument applies.

(11.7) can be then written

$$(11.8) \qquad \alpha p = x\zeta^{-g} + y\zeta^{1-g} - x\zeta^g - y\zeta^{g-1},$$

where α is an integer in K and none of the four exponents of ζ is divisible by p. The numbers $\zeta, \zeta^2, \ldots, \zeta^{p-1}$ form an integral basis for K, and the numbers $\zeta^{-g}, \zeta^{1-g}, \zeta^g, \zeta^{g-1}$ occur among them. Now

$$\alpha = \frac{x}{p}\zeta^{-g} + \frac{y}{p}\zeta^{1-g} - \frac{x}{p}\zeta^g - \frac{y}{p}\zeta^{g-1}.$$

If no two of the exponents are congruent modulo p then $p \mid x$ and $p \mid y$, for α is an integer and its representation in terms of the integral basis is unique and involves integral coefficients only. Since in fact $p \nmid x$, $p \nmid y$ by hypothesis, two of the exponents must be congruent modulo p. Since $g \not\equiv 0$, $g \not\equiv 1$, the only remaining possibility is $2g \equiv 1 \pmod{p}$.

Because $2g \equiv 1 \pmod{p}$, and $\zeta^p = 1$, (11.8) can be written

$$\alpha p\zeta^g = x + y\zeta - x\zeta^{2g} - y\zeta^{2g-1}$$

$$= (x - y)(1 - \zeta) = (x - y)\lambda.$$

Hence

$$N\alpha \cdot Np = N(x - y)N\lambda, \qquad N\alpha \cdot p^{p-1} = (x - y)^{p-1}p.$$

We conclude that $p \mid (x - y)$ —that is, $x \equiv y \pmod{p}$.

If we go back to the very beginning and write (11.4) as

$$\prod_{m=0}^{p-1} (x + \zeta^m z) = -y^p,$$

a similar argument shows that $x \equiv z$ (mod p). Hence

$$0 = x^p + y^p + z^p \equiv x^p + x^p + x^p \equiv 3x^p \quad (\text{mod } p).$$

Then $p \mid 3x^p$, but $p \nmid x$. Hence $p = 3$.

The only possible regular prime for which (11.4) has a solution is $p = 3$, and we shall rule out this case by showing that $x^3 + y^3 + z^3 = 0$ cannot have a solution in rational integers if $3 \nmid x$, $3 \nmid y$, $3 \nmid z$. Since $-1, 0, 1$ forms a complete residue system modulo 3 and $3 \nmid x$, $x \equiv \pm 1$ (mod 3). Hence

$$x = 3k \pm 1, \quad x^3 = 27k^3 \pm 27k^2 + 9k \pm 1,$$

so $x^3 \equiv \pm 1$ (mod 9). Similarly $y^3 \equiv \pm 1$, $z^3 \equiv \pm 1$ (mod 9), so that

$$0 = x^3 + y^3 + z^3 \equiv \pm 1 \pm 1 \pm 1 \quad (\text{mod } 9).$$

Obviously this can not be true for *any* choice of the $\pm$ signs. Theorem 11.10 is proved.

Problems

1. If (x, y, z) is a solution of $x^2 + y^2 = z^2$ in positive integers, show that $x^2 \geq 2y + 1$. Deduce that there are at most a finite number of solutions of the equation for any given value of x.
2. (a) Find all positive primitive solutions of $x^2 + y^2 = z^2$ with $y = 60$.
   (b) Find all positive solutions of $x^2 + y^2 = z^2$ with $y = 60$.
3. Show that a conjugate of a root of unity is also a root of unity.
4. Let $P(x)$ be a monic irreducible polynomial over J. Suppose each root of $P(x)$ has modulus at most 1. Prove that the roots of $P(x)$ are all roots of unity. Hint. Show that there exists no root of modulus less than 1.

5. Suppose there existed a solution of $x^n + y^n = z^n$ in positive integers x, y, z for some $n > 2$. Show that $x^n > ny^{n-1}$, $y^n > nx^{n-1}$, and thus $x > n$, $y > n$.

6. Where in Theorem 11.10 was use made of the condition that p is regular?

REFERENCES

1. Birkhoff and MacLane, *Survey of Modern Algebra*, Revised Ed., Macmillan, New York, 1953.
2. Borevich and Shafarevich, *Number Theory* (trans. N. Greenleaf), Academic Press, New York, 1966.
3. A. O. Gelfond, *Transcendental and Algebraic Numbers* (trans. L. F. Boron), Dover, New York, 1960.
4. Hardy and Wright, *The Theory of Numbers*, 4th Ed., Oxford U. Press, London, 1959.
5. E. Hecke, *Theorie der algebraischen Zahlen*, Akademische Verlagsgesellschaft, Leipzig, 1923. Reprinted by Chelsea, New York, 1970.
6. E. Landau, *Vorlesungen über Zahlentheorie*, 3 volumes, S. Hirzel, Leipzig, 1927. Reprinted by Chelsea, New York, 1969.
7. S. Lang, *Algebraic Number Theory*, Addison-Wesley, Reading, Mass., 1970.
8. W. J. LeVeque, *Topics in Number Theory*, Vol. II, Addison-Wesley, Reading, Mass., 1956.
9. R. Narasimhan et al., *Algebraic Number Theory*, Math. Pamphlets 4, Tata Institute, Bombay, 1966.
10. O. Ore, *Les corps Algébriques et la Théorie des Idéaux*, Gauthier-Villars, Paris, 1934.
11. L. W. Reid, *The Elements of the Theory of Algebraic Numbers*, Macmillan, New York, 1910.
12. P. Samuel, *Algebraic Theory of Numbers* (Trans. A. J. Silberger), Kershaw, London, 1972.
13. H. M. Stark, *An Introduction to Number Theory*, Markham, Chicago, 1970.
14. J. M. Thomas, *Theory of Equations*, McGraw-Hill, New York, 1938.
15. H. Weyl, *Algebraic Theory of Numbers*, Princeton U. Press, Princeton, 1940.

LIST OF SYMBOLS
with page on which defined

INDEX

A CATALOG OF SELECTED
DOVER BOOKS
IN SCIENCE AND MATHEMATICS

Astronomy

BURNHAM'S CELESTIAL HANDBOOK, Robert Burnham, Jr. Thorough guide to the stars beyond our solar system. Exhaustive treatment. Alphabetical by constellation: Andromeda to Cetus in Vol. 1; Chamaeleon to Orion in Vol. 2; and Pavo to Vulpecula in Vol. 3. Hundreds of illustrations. Index in Vol. 3. 2,000pp. 6⅛ x 9¼.

Vol. I: 0-486-23567-X
Vol. II: 0-486-23568-8
Vol. III: 0-486-23673-0

EXPLORING THE MOON THROUGH BINOCULARS AND SMALL TELE-SCOPES, Ernest H. Cherrington, Jr. Informative, profusely illustrated guide to locating and identifying craters, rills, seas, mountains, other lunar features. Newly revised and updated with special section of new photos. Over 100 photos and diagrams. 240pp. 8¼ x 11.   0-486-24491-1

THE EXTRATERRESTRIAL LIFE DEBATE, 1750–1900, Michael J. Crowe. First detailed, scholarly study in English of the many ideas that developed from 1750 to 1900 regarding the existence of intelligent extraterrestrial life. Examines ideas of Kant, Herschel, Voltaire, Percival Lowell, many other scientists and thinkers. 16 illustrations. 704pp. 5⅜ x 8½.   0-486-40675-X

THEORIES OF THE WORLD FROM ANTIQUITY TO THE COPERNICAN REVOLUTION, Michael J. Crowe. Newly revised edition of an accessible, enlightening book re-creates the change from an earth-centered to a sun-centered conception of the solar system. 242pp. 5⅜ x 8½.   0-486-41444-2

ARISTARCHUS OF SAMOS: The Ancient Copernicus, Sir Thomas Heath. Heath's history of astronomy ranges from Homer and Hesiod to Aristarchus and includes quotes from numerous thinkers, compilers, and scholasticists from Thales and Anaximander through Pythagoras, Plato, Aristotle, and Heraclides. 34 figures. 448pp. 5⅜ x 8½.
                                                                                    0-486-43886-4

A COMPLETE MANUAL OF AMATEUR ASTRONOMY: TOOLS AND TECHNIQUES FOR ASTRONOMICAL OBSERVATIONS, P. Clay Sherrod with Thomas L. Koed. Concise, highly readable book discusses: selecting, setting up and main-taining a telescope; amateur studies of the sun; lunar topography and occultations; obser-vations of Mars, Jupiter, Saturn, the minor planets and the stars; an introduction to pho-toelectric photometry; more. 1981 ed. 124 figures. 25 halftones. 37 tables. 335pp. 6½ x 9¼.   0-486-42820-8

AMATEUR ASTRONOMER'S HANDBOOK, J. B. Sidgwick. Timeless, comprehen-sive coverage of telescopes, mirrors, lenses, mountings, telescope drives, micrometers, spectroscopes, more. 189 illustrations. 576pp. 5⅝ x 8¼. (Available in U.S. only.)
                                                                                    0-486-24034-7

STAR LORE: Myths, Legends, and Facts, William Tyler Olcott. Captivating retellings of the origins and histories of ancient star groups include Pegasus, Ursa Major, Pleiades, signs of the zodiac, and other constellations. "Classic."—Sky & Telescope. 58 illustrations. 544pp. 5⅜ x 8½.   0-486-43581-4

Chemistry

THE SCEPTICAL CHYMIST: THE CLASSIC 1661 TEXT, Robert Boyle. Boyle defines the term "element," asserting that all natural phenomena can be explained by the motion and organization of primary particles. 1911 ed. viii+232pp. $5\frac{3}{8}$ x $8\frac{1}{2}$.
0-486-42825-7

RADIOACTIVE SUBSTANCES, Marie Curie. Here is the celebrated scientist's doctoral thesis, the prelude to her receipt of the 1903 Nobel Prize. Curie discusses establishing atomic character of radioactivity found in compounds of uranium and thorium; extraction from pitchblende of polonium and radium; isolation of pure radium chloride; determination of atomic weight of radium; plus electric, photographic, luminous, heat, color effects of radioactivity. ii+94pp. $5\frac{3}{8}$ x $8\frac{1}{2}$.
0-486-42550-9

CHEMICAL MAGIC, Leonard A. Ford. Second Edition, Revised by E. Winston Grundmeier. Over 100 unusual stunts demonstrating cold fire, dust explosions, much more. Text explains scientific principles and stresses safety precautions. 128pp. $5\frac{3}{8}$ x $8\frac{1}{2}$.
0-486-67628-5

MOLECULAR THEORY OF CAPILLARITY, J. S. Rowlinson and B. Widom. History of surface phenomena offers critical and detailed examination and assessment of modern theories, focusing on statistical mechanics and application of results in mean-field approximation to model systems. 1989 edition. 352pp. $5\frac{3}{8}$ x $8\frac{1}{2}$.
0-486-42544-4

CHEMICAL AND CATALYTIC REACTION ENGINEERING, James J. Carberry. Designed to offer background for managing chemical reactions, this text examines behavior of chemical reactions and reactors; fluid-fluid and fluid-solid reaction systems; heterogeneous catalysis and catalytic kinetics; more. 1976 edition. 672pp. $6\frac{1}{8}$ x $9\frac{1}{4}$.
0-486-41736-0 $31.95

ELEMENTS OF CHEMISTRY, Antoine Lavoisier. Monumental classic by founder of modern chemistry in remarkable reprint of rare 1790 Kerr translation. A must for every student of chemistry or the history of science. 539pp. $5\frac{3}{8}$ x $8\frac{1}{2}$.
0-486-64624-6

MOLECULES AND RADIATION: An Introduction to Modern Molecular Spectroscopy. Second Edition, Jeffrey I. Steinfeld. This unified treatment introduces upper-level undergraduates and graduate students to the concepts and the methods of molecular spectroscopy and applications to quantum electronics, lasers, and related optical phenomena. 1985 edition. 512pp. $5\frac{3}{8}$ x $8\frac{1}{2}$.
0-486-44152-0

A SHORT HISTORY OF CHEMISTRY, J. R. Partington. Classic exposition explores origins of chemistry, alchemy, early medical chemistry, nature of atmosphere, theory of valency, laws and structure of atomic theory, much more. 428pp. $5\frac{3}{8}$ x $8\frac{1}{2}$. (Available in U.S. only.)
0-486-65977-1

GENERAL CHEMISTRY, Linus Pauling. Revised 3rd edition of classic first-year text by Nobel laureate. Atomic and molecular structure, quantum mechanics, statistical mechanics, thermodynamics correlated with descriptive chemistry. Problems. 992pp. $5\frac{3}{8}$ x $8\frac{1}{2}$.
0-486-65622-5

ELECTRON CORRELATION IN MOLECULES, S. Wilson. This text addresses one of theoretical chemistry's central problems. Topics include molecular electronic structure, independent electron models, electron correlation, the linked diagram theorem, and related topics. 1984 edition. 304pp. $5\frac{3}{8}$ x $8\frac{1}{2}$.
0-486-45879-2

Engineering

DE RE METALLICA, Georgius Agricola. The famous Hoover translation of greatest treatise on technological chemistry, engineering, geology, mining of early modern times (1556). All 289 original woodcuts. 638pp. 6³/₄ x 11. 0-486-60006-8

FUNDAMENTALS OF ASTRODYNAMICS, Roger Bate et al. Modern approach developed by U.S. Air Force Academy. Designed as a first course. Problems, exercises. Numerous illustrations. 455pp. 5³/₈ x 8¹/₂. 0-486-60061-0

DYNAMICS OF FLUIDS IN POROUS MEDIA, Jacob Bear. For advanced students of ground water hydrology, soil mechanics and physics, drainage and irrigation engineering and more. 335 illustrations. Exercises, with answers. 784pp. 6¹/₈ x 9¹/₄. 0-486-65675-6

THEORY OF VISCOELASTICITY (SECOND EDITION), Richard M. Christensen. Complete consistent description of the linear theory of the viscoelastic behavior of materials. Problem-solving techniques discussed. 1982 edition. 29 figures. xiv+364pp. 6¹/₈ x 9¹/₄.
0-486-42880-X

MECHANICS, J. P. Den Hartog. A classic introductory text or refresher. Hundreds of applications and design problems illuminate fundamentals of trusses, loaded beams and cables, etc. 334 answered problems. 462pp. 5³/₈ x 8¹/₂. 0-486-60754-2

MECHANICAL VIBRATIONS, J. P. Den Hartog. Classic textbook offers lucid explanations and illustrative models, applying theories of vibrations to a variety of practical industrial engineering problems. Numerous figures. 233 problems, solutions. Appendix. Index. Preface. 436pp. 5³/₈ x 8¹/₂. 0-486-64785-4

STRENGTH OF MATERIALS, J. P. Den Hartog. Full, clear treatment of basic material (tension, torsion, bending, etc.) plus advanced material on engineering methods, applications. 350 answered problems. 323pp. 5³/₈ x 8¹/₂. 0-486-60755-0

A HISTORY OF MECHANICS, René Dugas. Monumental study of mechanical principles from antiquity to quantum mechanics. Contributions of ancient Greeks, Galileo, Leonardo, Kepler, Lagrange, many others. 671pp. 5³/₈ x 8¹/₂. 0-486-65632-2

STABILITY THEORY AND ITS APPLICATIONS TO STRUCTURAL MECHANICS, Clive L. Dym. Self-contained text focuses on Koiter postbuckling analyses, with mathematical notions of stability of motion. Basing minimum energy principles for static stability upon dynamic concepts of stability of motion, it develops asymptotic buckling and postbuckling analyses from potential energy considerations, with applications to columns, plates, and arches. 1974 ed. 208pp. 5³/₈ x 8¹/₂. 0-486-42541-X

BASIC ELECTRICITY, U.S. Bureau of Naval Personnel. Originally a training course; best nontechnical coverage. Topics include batteries, circuits, conductors, AC and DC, inductance and capacitance, generators, motors, transformers, amplifiers, etc. Many questions with answers. 349 illustrations. 1969 edition. 448pp. 6¹/₂ x 9¹/₄. 0-486-20973-3

ROCKETS, Robert Goddard. Two of the most significant publications in the history of rocketry and jet propulsion: "A Method of Reaching Extreme Altitudes" (1919) and "Liquid Propellant Rocket Development" (1936). 128pp. $5^3/_8$ x $8^1/_2$. 0-486-42537-1

STATISTICAL MECHANICS: PRINCIPLES AND APPLICATIONS, Terrell L. Hill. Standard text covers fundamentals of statistical mechanics, applications to fluctuation theory, imperfect gases, distribution functions, more. 448pp. $5^3/_8$ x $8^1/_2$. 0-486-65390-0

ENGINEERING AND TECHNOLOGY 1650–1750: ILLUSTRATIONS AND TEXTS FROM ORIGINAL SOURCES, Martin Jensen. Highly readable text with more than 200 contemporary drawings and detailed engravings of engineering projects dealing with surveying, leveling, materials, hand tools, lifting equipment, transport and erection, piling, bailing, water supply, hydraulic engineering, and more. Among the specific projects outlined-transporting a 50-ton stone to the Louvre, erecting an obelisk, building timber locks, and dredging canals. 207pp. $8^3/_8$ x $11^1/_4$. 0-486-42232-1

THE VARIATIONAL PRINCIPLES OF MECHANICS, Cornelius Lanczos. Graduate level coverage of calculus of variations, equations of motion, relativistic mechanics, more. First inexpensive paperbound edition of classic treatise. Index. Bibliography. 418pp. $5^3/_8$ x $8^1/_2$. 0-486-65067-7

PROTECTION OF ELECTRONIC CIRCUITS FROM OVERVOLTAGES, Ronald B. Standler. Five-part treatment presents practical rules and strategies for circuits designed to protect electronic systems from damage by transient overvoltages. 1989 ed. xxiv+434pp. $6^1/_8$ x $9^1/_4$. 0-486-42552-5

ROTARY WING AERODYNAMICS, W. Z. Stepniewski. Clear, concise text covers aerodynamic phenomena of the rotor and offers guidelines for helicopter performance evaluation. Originally prepared for NASA. 537 figures. 640pp. $6^1/_8$ x $9^1/_4$. 0-486-64647-5

INTRODUCTION TO SPACE DYNAMICS, William Tyrrell Thomson. Comprehensive, classic introduction to space-flight engineering for advanced undergraduate and graduate students. Includes vector algebra, kinematics, transformation of coordinates. Bibliography. Index. 352pp. $5^3/_8$ x $8^1/_2$. 0-486-65113-4

HISTORY OF STRENGTH OF MATERIALS, Stephen P. Timoshenko. Excellent historical survey of the strength of materials with many references to the theories of elasticity and structure. 245 figures. 452pp. $5^3/_8$ x $8^1/_2$. 0-486-61187-6

ANALYTICAL FRACTURE MECHANICS, David J. Unger. Self-contained text supplements standard fracture mechanics texts by focusing on analytical methods for determining crack-tip stress and strain fields. 336pp. $6^1/_8$ x $9^1/_4$. 0-486-41737-9

STATISTICAL MECHANICS OF ELASTICITY, J. H. Weiner. Advanced, self-contained treatment illustrates general principles and elastic behavior of solids. Part 1, based on classical mechanics, studies thermoelastic behavior of crystalline and polymeric solids. Part 2, based on quantum mechanics, focuses on interatomic force laws, behavior of solids, and thermally activated processes. For students of physics and chemistry and for polymer physicists. 1983 ed. 96 figures. 496pp. $5^3/_8$ x $8^1/_2$. 0-486-42260-7

Mathematics

FUNCTIONAL ANALYSIS (Second Corrected Edition), George Bachman and Lawrence Narici. Excellent treatment of subject geared toward students with background in linear algebra, advanced calculus, physics and engineering. Text covers introduction to inner-product spaces, normed, metric spaces, and topological spaces; complete orthonormal sets, the Hahn-Banach Theorem and its consequences, and many other related subjects. 1966 ed. 544pp. 6⅛ x 9¼. 0-486-40251-7

DIFFERENTIAL MANIFOLDS, Antoni A. Kosinski. Introductory text for advanced undergraduates and graduate students presents systematic study of the topological structure of smooth manifolds, starting with elements of theory and concluding with method of surgery. 1993 edition. 288pp. 5⅜ x 8½. 0-486-46244-7

VECTOR AND TENSOR ANALYSIS WITH APPLICATIONS, A. I. Borisenko and I. E. Tarapov. Concise introduction. Worked-out problems, solutions, exercises. 257pp. 5⅝ x 8¼. 0-486-63833-2

AN INTRODUCTION TO ORDINARY DIFFERENTIAL EQUATIONS, Earl A. Coddington. A thorough and systematic first course in elementary differential equations for undergraduates in mathematics and science, with many exercises and problems (with answers). Index. 304pp. 5⅜ x 8½. 0-486-65942-9

FOURIER SERIES AND ORTHOGONAL FUNCTIONS, Harry F. Davis. An incisive text combining theory and practical example to introduce Fourier series, orthogonal functions and applications of the Fourier method to boundary-value problems. 570 exercises. Answers and notes. 416pp. 5⅜ x 8½. 0-486-65973-9

COMPUTABILITY AND UNSOLVABILITY, Martin Davis. Classic graduate-level introduction to theory of computability, usually referred to as theory of recurrent functions. New preface and appendix. 288pp. 5⅜ x 8½. 0-486-61471-9

AN INTRODUCTION TO MATHEMATICAL ANALYSIS, Robert A. Rankin. Dealing chiefly with functions of a single real variable, this text by a distinguished educator introduces limits, continuity, differentiability, integration, convergence of infinite series, double series, and infinite products. 1963 edition. 624pp. 5⅜ x 8½. 0-486-46251-X

METHODS OF NUMERICAL INTEGRATION (SECOND EDITION), Philip J. Davis and Philip Rabinowitz. Requiring only a background in calculus, this text covers approximate integration over finite and infinite intervals, error analysis, approximate integration in two or more dimensions, and automatic integration. 1984 edition. 624pp. 5⅜ x 8½. 0-486-45339-1

INTRODUCTION TO LINEAR ALGEBRA AND DIFFERENTIAL EQUATIONS, John W. Dettman. Excellent text covers complex numbers, determinants, orthonormal bases, Laplace transforms, much more. Exercises with solutions. Undergraduate level. 416pp. 5⅜ x 8½. 0-486-65191-6

RIEMANN'S ZETA FUNCTION, H. M. Edwards. Superb, high-level study of landmark 1859 publication entitled "On the Number of Primes Less Than a Given Magnitude" traces developments in mathematical theory that it inspired. xiv+315pp. 5⅜ x 8½. 0-486-41740-9

CALCULUS OF VARIATIONS WITH APPLICATIONS, George M. Ewing. Applications-oriented introduction to variational theory develops insight and promotes understanding of specialized books, research papers. Suitable for advanced undergraduate/graduate students as primary, supplementary text. 352pp. 5⅜ x 8½.
0-486-64856-7

MATHEMATICIAN'S DELIGHT, W. W. Sawyer. "Recommended with confidence" by *The Times Literary Supplement,* this lively survey was written by a renowned teacher. It starts with arithmetic and algebra, gradually proceeding to trigonometry and calculus. 1943 edition. 240pp. 5⅜ x 8½.
0-486-46240-4

ADVANCED EUCLIDEAN GEOMETRY, Roger A. Johnson. This classic text explores the geometry of the triangle and the circle, concentrating on extensions of Euclidean theory, and examining in detail many relatively recent theorems. 1929 edition. 336pp. 5⅜ x 8½.
0-486-46237-4

COUNTEREXAMPLES IN ANALYSIS, Bernard R. Gelbaum and John M. H. Olmsted. These counterexamples deal mostly with the part of analysis known as "real variables." The first half covers the real number system, and the second half encompasses higher dimensions. 1962 edition. xxiv+198pp. 5⅜ x 8½.
0-486-42875-3

CATASTROPHE THEORY FOR SCIENTISTS AND ENGINEERS, Robert Gilmore. Advanced-level treatment describes mathematics of theory grounded in the work of Poincaré, R. Thom, other mathematicians. Also important applications to problems in mathematics, physics, chemistry and engineering. 1981 edition. References. 28 tables. 397 black-and-white illustrations. xvii + 666pp. 6⅛ x 9¼.
0-486-67539-4

COMPLEX VARIABLES: Second Edition, Robert B. Ash and W. P. Novinger. Suitable for advanced undergraduates and graduate students, this newly revised treatment covers Cauchy theorem and its applications, analytic functions, and the prime number theorem. Numerous problems and solutions. 2004 edition. 224pp. 6½ x 9¼.
0-486-46250-1

NUMERICAL METHODS FOR SCIENTISTS AND ENGINEERS, Richard Hamming. Classic text stresses frequency approach in coverage of algorithms, polynomial approximation, Fourier approximation, exponential approximation, other topics. Revised and enlarged 2nd edition. 721pp. 5⅜ x 8½.
0-486-65241-6

INTRODUCTION TO NUMERICAL ANALYSIS (2nd Edition), F. B. Hildebrand. Classic, fundamental treatment covers computation, approximation, interpolation, numerical differentiation and integration, other topics. 150 new problems. 669pp. 5⅜ x 8½.
0-486-65363-3

MARKOV PROCESSES AND POTENTIAL THEORY, Robert M. Blumental and Ronald K. Getoor. This graduate-level text explores the relationship between Markov processes and potential theory in terms of excessive functions, multiplicative functionals and subprocesses, additive functionals and their potentials, and dual processes. 1968 edition. 320pp. 5⅜ x 8½.
0-486-46263-3

ABSTRACT SETS AND FINITE ORDINALS: An Introduction to the Study of Set Theory, G. B. Keene. This text unites logical and philosophical aspects of set theory in a manner intelligible to mathematicians without training in formal logic and to logicians without a mathematical background. 1961 edition. 112pp. 5⅜ x 8½.
0-486-46249-8

INTRODUCTORY REAL ANALYSIS, A.N. Kolmogorov, S. V. Fomin. Translated by Richard A. Silverman. Self-contained, evenly paced introduction to real and functional analysis. Some 350 problems. 403pp. 5⅜ x 8½.   0-486-61226-0

APPLIED ANALYSIS, Cornelius Lanczos. Classic work on analysis and design of finite processes for approximating solution of analytical problems. Algebraic equations, matrices, harmonic analysis, quadrature methods, much more. 559pp. 5⅜ x 8½.   0-486-65656-X

AN INTRODUCTION TO ALGEBRAIC STRUCTURES, Joseph Landin. Superb self-contained text covers "abstract algebra": sets and numbers, theory of groups, theory of rings, much more. Numerous well-chosen examples, exercises. 247pp. 5⅜ x 8½.
0-486-65940-2

QUALITATIVE THEORY OF DIFFERENTIAL EQUATIONS, V. V. Nemytskii and V.V. Stepanov. Classic graduate-level text by two prominent Soviet mathematicians covers classical differential equations as well as topological dynamics and ergodic theory. Bibliographies. 523pp. 5⅜ x 8½.   0-486-65954-2

THEORY OF MATRICES, Sam Perlis. Outstanding text covering rank, nonsingularity and inverses in connection with the development of canonical matrices under the relation of equivalence, and without the intervention of determinants. Includes exercises. 237pp. 5⅜ x 8½.   0-486-66810-X

INTRODUCTION TO ANALYSIS, Maxwell Rosenlicht. Unusually clear, accessible coverage of set theory, real number system, metric spaces, continuous functions, Riemann integration, multiple integrals, more. Wide range of problems. Undergraduate level. Bibliography. 254pp. 5⅜ x 8½.   0-486-65038-3

MODERN NONLINEAR EQUATIONS, Thomas L. Saaty. Emphasizes practical solution of problems; covers seven types of equations. ". . . a welcome contribution to the existing literature. . . ."—*Math Reviews.* 490pp. 5⅜ x 8½.   0-486-64232-1

MATRICES AND LINEAR ALGEBRA, Hans Schneider and George Phillip Barker. Basic textbook covers theory of matrices and its applications to systems of linear equations and related topics such as determinants, eigenvalues and differential equations. Numerous exercises. 432pp. 5⅜ x 8½.   0-486-66014-1

LINEAR ALGEBRA, Georgi E. Shilov. Determinants, linear spaces, matrix algebras, similar topics. For advanced undergraduates, graduates. Silverman translation. 387pp. 5⅜ x 8½.   0-486-63518-X

MATHEMATICAL METHODS OF GAME AND ECONOMIC THEORY: Revised Edition, Jean-Pierre Aubin. This text begins with optimization theory and convex analysis, followed by topics in game theory and mathematical economics, and concluding with an introduction to nonlinear analysis and control theory. 1982 edition. 656pp. 6⅛ x 9¼.
0-486-46265-X

SET THEORY AND LOGIC, Robert R. Stoll. Lucid introduction to unified theory of mathematical concepts. Set theory and logic seen as tools for conceptual understanding of real number system. 496pp. 5⅜ x 8¼.   0-486-63829-4

TENSOR CALCULUS, J.L. Synge and A. Schild. Widely used introductory text covers spaces and tensors, basic operations in Riemannian space, non-Riemannian spaces, etc. 324pp. $5^3/_8$ x $8^1/_4$. 0-486-63612-7

ORDINARY DIFFERENTIAL EQUATIONS, Morris Tenenbaum and Harry Pollard. Exhaustive survey of ordinary differential equations for undergraduates in mathematics, engineering, science. Thorough analysis of theorems. Diagrams. Bibliography. Index. 818pp. $5^3/_8$ x $8^1/_2$. 0-486-64940-7

INTEGRAL EQUATIONS, F. G. Tricomi. Authoritative, well-written treatment of extremely useful mathematical tool with wide applications. Volterra Equations, Fredholm Equations, much more. Advanced undergraduate to graduate level. Exercises. Bibliography. 238pp. $5^3/_8$ x $8^1/_2$. 0-486-64828-1

FOURIER SERIES, Georgi P. Tolstov. Translated by Richard A. Silverman. A valuable addition to the literature on the subject, moving clearly from subject to subject and theorem to theorem. 107 problems, answers. 336pp. $5^3/_8$ x $8^1/_2$. 0-486-63317-9

INTRODUCTION TO MATHEMATICAL THINKING, Friedrich Waismann. Examinations of arithmetic, geometry, and theory of integers; rational and natural numbers; complete induction; limit and point of accumulation; remarkable curves; complex and hypercomplex numbers, more. 1959 ed. 27 figures. xii+260pp. $5^3/_8$ x $8^1/_2$. 0-486-42804-8

THE RADON TRANSFORM AND SOME OF ITS APPLICATIONS, Stanley R. Deans. Of value to mathematicians, physicists, and engineers, this excellent introduction covers both theory and applications, including a rich array of examples and literature. Revised and updated by the author. 1993 edition. 304pp. $6^1/_8$ x $9^1/_4$. 0-486-46241-2

CALCULUS OF VARIATIONS, Robert Weinstock. Basic introduction covering isoperimetric problems, theory of elasticity, quantum mechanics, electrostatics, etc. Exercises throughout. 326pp. $5^3/_8$ x $8^1/_2$. 0-486-63069-2

THE CONTINUUM: A CRITICAL EXAMINATION OF THE FOUNDATION OF ANALYSIS, Hermann Weyl. Classic of 20th-century foundational research deals with the conceptual problem posed by the continuum. 156pp. $5^3/_8$ x $8^1/_2$. 0-486-67982-9

CHALLENGING MATHEMATICAL PROBLEMS WITH ELEMENTARY SOLUTIONS, A. M. Yaglom and I. M. Yaglom. Over 170 challenging problems on probability theory, combinatorial analysis, points and lines, topology, convex polygons, many other topics. Solutions. Total of 445pp. $5^3/_8$ x $8^1/_2$. Two-vol. set.
Vol. I: 0-486-65536-9 Vol. II: 0-486-65537-7

INTRODUCTION TO PARTIAL DIFFERENTIAL EQUATIONS WITH APPLICATIONS, E. C. Zachmanoglou and Dale W. Thoe. Essentials of partial differential equations applied to common problems in engineering and the physical sciences. Problems and answers. 416pp. $5^3/_8$ x $8^1/_2$. 0-486-65251-3

STOCHASTIC PROCESSES AND FILTERING THEORY, Andrew H. Jazwinski. This unified treatment presents material previously available only in journals, and in terms accessible to engineering students. Although theory is emphasized, it discusses numerous practical applications as well. 1970 edition. 400pp. $5^3/_8$ x $8^1/_2$. 0-486-46274-9

Math—Decision Theory, Statistics, Probability

INTRODUCTION TO PROBABILITY, John E. Freund. Featured topics include permutations and factorials, probabilities and odds, frequency interpretation, mathematical expectation, decision-making, postulates of probability, rule of elimination, much more. Exercises with some solutions. Summary. 1973 edition. 247pp. 5³/₈ x 8¹/₂.
0-486-67549-1

STATISTICAL AND INDUCTIVE PROBABILITIES, Hugues Leblanc. This treatment addresses a decades-old dispute among probability theorists, asserting that both statistical and inductive probabilities may be treated as sentence-theoretic measurements, and that the latter qualify as estimates of the former. 1962 edition. 160pp. 5³/₈ x 8¹/₂.
0-486-44980-7

APPLIED MULTIVARIATE ANALYSIS: Using Bayesian and Frequentist Methods of Inference, Second Edition, S. James Press. This two-part treatment deals with foundations as well as models and applications. Topics include continuous multivariate distributions; regression and analysis of variance; factor analysis and latent structure analysis; and structuring multivariate populations. 1982 edition. 692pp. 5³/₈ x 8¹/₂.   0-486-44236-5

LINEAR PROGRAMMING AND ECONOMIC ANALYSIS, Robert Dorfman, Paul A. Samuelson and Robert M. Solow. First comprehensive treatment of linear programming in standard economic analysis. Game theory, modern welfare economics, Leontief input-output, more. 525pp. 5³/₈ x 8¹/₂.   0-486-65491-5

PROBABILITY: AN INTRODUCTION, Samuel Goldberg. Excellent basic text covers set theory, probability theory for finite sample spaces, binomial theorem, much more. 360 problems. Bibliographies. 322pp. 5³/₈ x 8¹/₂.   0-486-65252-1

GAMES AND DECISIONS: INTRODUCTION AND CRITICAL SURVEY, R. Duncan Luce and Howard Raiffa. Superb nontechnical introduction to game theory, primarily applied to social sciences. Utility theory, zero-sum games, n-person games, decision-making, much more. Bibliography. 509pp. 5³/₈ x 8¹/₂.   0-486-65943-7

INTRODUCTION TO THE THEORY OF GAMES, J. C. C. McKinsey. This comprehensive overview of the mathematical theory of games illustrates applications to situations involving conflicts of interest, including economic, social, political, and military contexts. Appropriate for advanced undergraduate and graduate courses; advanced calculus a prerequisite. 1952 ed. x+372pp. 5³/₈ x 8¹/₂.   0-486-42811-7

FIFTY CHALLENGING PROBLEMS IN PROBABILITY WITH SOLUTIONS, Frederick Mosteller. Remarkable puzzlers, graded in difficulty, illustrate elementary and advanced aspects of probability. Detailed solutions. 88pp. 5³/₈ x 8¹/₂.   0-486-65355-2

PROBABILITY THEORY: A CONCISE COURSE, Y. A. Rozanov. Highly readable, self-contained introduction covers combination of events, dependent events, Bernoulli trials, etc. 148pp. 5⁵/₈ x 8¹/₄.   0-486-63544-9

THE STATISTICAL ANALYSIS OF EXPERIMENTAL DATA, John Mandel. First half of book presents fundamental mathematical definitions, concepts and facts while remaining half deals with statistics primarily as an interpretive tool. Well-written text, numerous worked examples with step-by-step presentation. Includes 116 tables. 448pp. 5³/₈ x 8¹/₂.   0-486-64666-1

Math—Geometry and Topology

ELEMENTARY CONCEPTS OF TOPOLOGY, Paul Alexandroff. Elegant, intuitive approach to topology from set-theoretic topology to Betti groups; how concepts of topology are useful in math and physics. 25 figures. 57pp. 5⅜ x 8½.   0-486-60747-X

A LONG WAY FROM EUCLID, Constance Reid. Lively guide by a prominent historian focuses on the role of Euclid's Elements in subsequent mathematical developments. Elementary algebra and plane geometry are sole prerequisites. 80 drawings. 1963 edition. 304pp. 5⅜ x 8½.   0-486-43613-6

EXPERIMENTS IN TOPOLOGY, Stephen Barr. Classic, lively explanation of one of the byways of mathematics. Klein bottles, Moebius strips, projective planes, map coloring, problem of the Koenigsberg bridges, much more, described with clarity and wit. 43 figures. 210pp. 5⅜ x 8½.   0-486-25933-1

THE GEOMETRY OF RENÉ DESCARTES, René Descartes. The great work founded analytical geometry. Original French text, Descartes's own diagrams, together with definitive Smith-Latham translation. 244pp. 5⅜ x 8½.   0-486-60068-8

EUCLIDEAN GEOMETRY AND TRANSFORMATIONS, Clayton W. Dodge. This introduction to Euclidean geometry emphasizes transformations, particularly isometries and similarities. Suitable for undergraduate courses, it includes numerous examples, many with detailed answers. 1972 ed. viii+296pp. 6⅛ x 9¼.   0-486-43476-1

EXCURSIONS IN GEOMETRY, C. Stanley Ogilvy. A straightedge, compass, and a little thought are all that's needed to discover the intellectual excitement of geometry. Harmonic division and Apollonian circles, inversive geometry, hexlet, Golden Section, more. 132 illustrations. 192pp. 5⅜ x 8½.   0-486-26530-7

THE THIRTEEN BOOKS OF EUCLID'S ELEMENTS, translated with introduction and commentary by Sir Thomas L. Heath. Definitive edition. Textual and linguistic notes, mathematical analysis. 2,500 years of critical commentary. Unabridged. 1,414pp. 5⅜ x 8½. Three-vol. set.
          Vol. I: 0-486-60088-2   Vol. II: 0-486-60089-0   Vol. III: 0-486-60090-4

SPACE AND GEOMETRY: IN THE LIGHT OF PHYSIOLOGICAL, PSYCHOLOGICAL AND PHYSICAL INQUIRY, Ernst Mach. Three essays by an eminent philosopher and scientist explore the nature, origin, and development of our concepts of space, with a distinctness and precision suitable for undergraduate students and other readers. 1906 ed. vi+148pp. 5⅜ x 8½.   0-486-43909-7

GEOMETRY OF COMPLEX NUMBERS, Hans Schwerdtfeger. Illuminating, widely praised book on analytic geometry of circles, the Moebius transformation, and two-dimensional non-Euclidean geometries. 200pp. 5⅝ x 8¼.   0-486-63830-8

DIFFERENTIAL GEOMETRY, Heinrich W. Guggenheimer. Local differential geometry as an application of advanced calculus and linear algebra. Curvature, transformation groups, surfaces, more. Exercises. 62 figures. 378pp. 5⅜ x 8½.   0-486-63433-7

History of Math

THE WORKS OF ARCHIMEDES, Archimedes (T. L. Heath, ed.). Topics include the famous problems of the ratio of the areas of a cylinder and an inscribed sphere; the measurement of a circle; the properties of conoids, spheroids, and spirals; and the quadrature of the parabola. Informative introduction. clxxxvi+326pp. 5³/₈ x 8¹/₂.   0-486-42084-1

A SHORT ACCOUNT OF THE HISTORY OF MATHEMATICS, W. W. Rouse Ball. One of clearest, most authoritative surveys from the Egyptians and Phoenicians through 19th-century figures such as Grassman, Galois, Riemann. Fourth edition. 522pp. 5³/₈ x 8¹/₂.   0-486-20630-0

THE HISTORY OF THE CALCULUS AND ITS CONCEPTUAL DEVELOP-MENT, Carl B. Boyer. Origins in antiquity, medieval contributions, work of Newton, Leibniz, rigorous formulation. Treatment is verbal. 346pp. 5³/₈ x 8¹/₂.   0-486-60509-4

THE HISTORICAL ROOTS OF ELEMENTARY MATHEMATICS, Lucas N. H. Bunt, Phillip S. Jones, and Jack D. Bedient. Fundamental underpinnings of modern arithmetic, algebra, geometry and number systems derived from ancient civilizations. 320pp. 5³/₈ x 8¹/₂.   0-486-25563-8

THE HISTORY OF THE CALCULUS AND ITS CONCEPTUAL DEVELOP-MENT, Carl B. Boyer. Fluent description of the development of both the integral and differential calculus—its early beginnings in antiquity, medieval contributions, and a consideration of Newton and Leibniz. 368pp. 5³/₈ x 8¹/₂.   0-486-60509-4

GAMES, GODS & GAMBLING: A HISTORY OF PROBABILITY AND STATISTICAL IDEAS, F. N. David. Episodes from the lives of Galileo, Fermat, Pascal, and others illustrate this fascinating account of the roots of mathematics. Features thought-provoking references to classics, archaeology, biography, poetry. 1962 edition. 304pp. 5³/₈ x 8¹/₂. (Available in U.S. only.)   0-486-40023-9

OF MEN AND NUMBERS: THE STORY OF THE GREAT MATHEMATICIANS, Jane Muir. Fascinating accounts of the lives and accomplishments of history's greatest mathematical minds—Pythagoras, Descartes, Euler, Pascal, Cantor, many more. Anecdotal, illuminating. 30 diagrams. Bibliography. 256pp. 5³/₈ x 8¹/₂.   0-486-28973-7

HISTORY OF MATHEMATICS, David E. Smith. Nontechnical survey from ancient Greece and Orient to late 19th century; evolution of arithmetic, geometry, trigonometry, calculating devices, algebra, the calculus. 362 illustrations. 1,355pp. 5³/₈ x 8¹/₂. Two-vol. set.   Vol. I: 0-486-20429-4   Vol. II: 0-486-20430-8

A CONCISE HISTORY OF MATHEMATICS, Dirk J. Struik. The best brief history of mathematics. Stresses origins and covers every major figure from ancient Near East to 19th century. 41 illustrations. 195pp. 5³/₈ x 8¹/₂.   0-486-60255-9

Physics

OPTICAL RESONANCE AND TWO-LEVEL ATOMS, L. Allen and J. H. Eberly. Clear, comprehensive introduction to basic principles behind all quantum optical resonance phenomena. 53 illustrations. Preface. Index. 256pp. 5⅜ x 8½.　　0-486-65533-4

QUANTUM THEORY, David Bohm. This advanced undergraduate-level text presents the quantum theory in terms of qualitative and imaginative concepts, followed by specific applications worked out in mathematical detail. Preface. Index. 655pp. 5⅜ x 8½.
0-486-65969-0

ATOMIC PHYSICS (8th EDITION), Max Born. Nobel laureate's lucid treatment of kinetic theory of gases, elementary particles, nuclear atom, wave-corpuscles, atomic structure and spectral lines, much more. Over 40 appendices, bibliography. 495pp. 5⅜ x 8½.
0-486-65984-4

A SOPHISTICATE'S PRIMER OF RELATIVITY, P. W. Bridgman. Geared toward readers already acquainted with special relativity, this book transcends the view of theory as a working tool to answer natural questions: What is a frame of reference? What is a "law of nature"? What is the role of the "observer"? Extensive treatment, written in terms accessible to those without a scientific background. 1983 ed. xlviii+172pp. 5⅜ x 8½.
0-486-42549-5

AN INTRODUCTION TO HAMILTONIAN OPTICS, H. A. Buchdahl. Detailed account of the Hamiltonian treatment of aberration theory in geometrical optics. Many classes of optical systems defined in terms of the symmetries they possess. Problems with detailed solutions. 1970 edition. xv + 360pp. 5⅜ x 8½.　　0-486-67597-1

PRIMER OF QUANTUM MECHANICS, Marvin Chester. Introductory text examines the classical quantum bead on a track: its state and representations; operator eigenvalues; harmonic oscillator and bound bead in a symmetric force field; and bead in a spherical shell. Other topics include spin, matrices, and the structure of quantum mechanics; the simplest atom; indistinguishable particles; and stationary-state perturbation theory. 1992 ed. xiv+314pp. 6⅛ x 9¼.　　0-486-42878-8

LECTURES ON QUANTUM MECHANICS, Paul A. M. Dirac. Four concise, brilliant lectures on mathematical methods in quantum mechanics from Nobel Prize-winning quantum pioneer build on idea of visualizing quantum theory through the use of classical mechanics. 96pp. 5⅜ x 8½.　　0-486-41713-1

THIRTY YEARS THAT SHOOK PHYSICS: THE STORY OF QUANTUM THEORY, George Gamow. Lucid, accessible introduction to influential theory of energy and matter. Careful explanations of Dirac's anti-particles, Bohr's model of the atom, much more. 12 plates. Numerous drawings. 240pp. 5⅜ x 8½.　　0-486-24895-X

ELECTRONIC STRUCTURE AND THE PROPERTIES OF SOLIDS: THE PHYSICS OF THE CHEMICAL BOND, Walter A. Harrison. Innovative text offers basic understanding of the electronic structure of covalent and ionic solids, simple metals, transition metals and their compounds. Problems. 1980 edition. 582pp. 6⅛ x 9¼.
0-486-66021-4

HYDRODYNAMIC AND HYDROMAGNETIC STABILITY, S. Chandrasekhar. Lucid examination of the Rayleigh-Benard problem; clear coverage of the theory of instabilities causing convection. 704pp. $5\frac{3}{8}$ x $8\frac{1}{4}$.　　　　　　　　0-486-64071-X

INVESTIGATIONS ON THE THEORY OF THE BROWNIAN MOVEMENT, Albert Einstein. Five papers (1905–8) investigating dynamics of Brownian motion and evolving elementary theory. Notes by R. Fürth. 122pp. $5\frac{3}{8}$ x $8\frac{1}{2}$.　　　0-486-60304-0

THE PHYSICS OF WAVES, William C. Elmore and Mark A. Heald. Unique overview of classical wave theory. Acoustics, optics, electromagnetic radiation, more. Ideal as classroom text or for self-study. Problems. 477pp. $5\frac{3}{8}$ x $8\frac{1}{2}$.　　　0-486-64926-1

GRAVITY, George Gamow. Distinguished physicist and teacher takes reader-friendly look at three scientists whose work unlocked many of the mysteries behind the laws of physics: Galileo, Newton, and Einstein. Most of the book focuses on Newton's ideas, with a concluding chapter on post-Einsteinian speculations concerning the relationship between gravity and other physical phenomena. 160pp. $5\frac{3}{8}$ x $8\frac{1}{2}$.　　　0-486-42563-0

PHYSICAL PRINCIPLES OF THE QUANTUM THEORY, Werner Heisenberg. Nobel Laureate discusses quantum theory, uncertainty, wave mechanics, work of Dirac, Schroedinger, Compton, Wilson, Einstein, etc. 184pp. $5\frac{3}{8}$ x $8\frac{1}{2}$.　　　0-486-60113-7

ATOMIC SPECTRA AND ATOMIC STRUCTURE, Gerhard Herzberg. One of best introductions; especially for specialist in other fields. Treatment is physical rather than mathematical. 80 illustrations. 257pp. $5\frac{3}{8}$ x $8\frac{1}{2}$.　　　0-486-60115-3

AN INTRODUCTION TO STATISTICAL THERMODYNAMICS, Terrell L. Hill. Excellent basic text offers wide-ranging coverage of quantum statistical mechanics, systems of interacting molecules, quantum statistics, more. 523pp. $5\frac{3}{8}$ x $8\frac{1}{2}$.　　　0-486-65242-4

THEORETICAL PHYSICS, Georg Joos, with Ira M. Freeman. Classic overview covers essential math, mechanics, electromagnetic theory, thermodynamics, quantum mechanics, nuclear physics, other topics. First paperback edition. xxiii + 885pp. $5\frac{3}{8}$ x $8\frac{1}{2}$.

0-486-65227-0

PROBLEMS AND SOLUTIONS IN QUANTUM CHEMISTRY AND PHYSICS, Charles S. Johnson, Jr. and Lee G. Pedersen. Unusually varied problems, detailed solutions in coverage of quantum mechanics, wave mechanics, angular momentum, molecular spectroscopy, more. 280 problems plus 139 supplementary exercises. 430pp. $6\frac{1}{2}$ x $9\frac{1}{4}$.

0-486-65236-X

THEORETICAL SOLID STATE PHYSICS, Vol. 1: Perfect Lattices in Equilibrium; Vol. II: Non-Equilibrium and Disorder, William Jones and Norman H. March. Monumental reference work covers fundamental theory of equilibrium properties of perfect crystalline solids, non-equilibrium properties, defects and disordered systems. Appendices. Problems. Preface. Diagrams. Index. Bibliography. Total of 1,301pp. $5\frac{3}{8}$ x $8\frac{1}{2}$. Two volumes.　　　Vol. I: 0-486-65015-4　Vol. II: 0-486-65016-2

WHAT IS RELATIVITY? L. D. Landau and G. B. Rumer. Written by a Nobel Prize physicist and his distinguished colleague, this compelling book explains the special theory of relativity to readers with no scientific background, using such familiar objects as trains, rulers, and clocks. 1960 ed. vi+72pp. $5\frac{3}{8}$ x $8\frac{1}{2}$.　　　0-486-42806-0

A TREATISE ON ELECTRICITY AND MAGNETISM, James Clerk Maxwell. Important foundation work of modern physics. Brings to final form Maxwell's theory of electromagnetism and rigorously derives his general equations of field theory. 1,084pp. $5^3/_8$ x $8^1/_2$. Two-vol. set. Vol. I: 0-486-60636-8 Vol. II: 0-486-60637-6

MATHEMATICS FOR PHYSICISTS, Philippe Dennery and Andre Krzywicki. Superb text provides math needed to understand today's more advanced topics in physics and engineering. Theory of functions of a complex variable, linear vector spaces, much more. Problems. 1967 edition. 400pp. $6^1/_2$ x $9^1/_4$. 0-486-69193-4

INTRODUCTION TO QUANTUM MECHANICS WITH APPLICATIONS TO CHEMISTRY, Linus Pauling & E. Bright Wilson, Jr. Classic undergraduate text by Nobel Prize winner applies quantum mechanics to chemical and physical problems. Numerous tables and figures enhance the text. Chapter bibliographies. Appendices. Index. 468pp. $5^3/_8$ x $8^1/_2$. 0-486-64871-0

METHODS OF THERMODYNAMICS, Howard Reiss. Outstanding text focuses on physical technique of thermodynamics, typical problem areas of understanding, and significance and use of thermodynamic potential. 1965 edition. 238pp. $5^3/_8$ x $8^1/_2$. 0-486-69445-3

THE ELECTROMAGNETIC FIELD, Albert Shadowitz. Comprehensive under- graduate text covers basics of electric and magnetic fields, builds up to electromagnetic theory. Also related topics, including relativity. Over 900 problems. 768pp. $5^3/_8$ x $8^1/_4$. 0-486-65660-8

GREAT EXPERIMENTS IN PHYSICS: FIRSTHAND ACCOUNTS FROM GALILEO TO EINSTEIN, Morris H. Shamos (ed.). 25 crucial discoveries: Newton's laws of motion, Chadwick's study of the neutron, Hertz on electromagnetic waves, more. Original accounts clearly annotated. 370pp. $5^3/_8$ x $8^1/_2$. 0-486-25346-5

EINSTEIN'S LEGACY, Julian Schwinger. A Nobel Laureate relates fascinating story of Einstein and development of relativity theory in well-illustrated, nontechnical volume. Subjects include meaning of time, paradoxes of space travel, gravity and its effect on light, non-Euclidean geometry and curving of space-time, impact of radio astronomy and space-age discoveries, and more. 189 b/w illustrations. xiv+250pp. $8^3/_8$ x $9^1/_4$. 0-486-41974-6

THE VARIATIONAL PRINCIPLES OF MECHANICS, Cornelius Lanczos. Philosophic, less formalistic approach to analytical mechanics offers model of clear, scholarly exposition at graduate level with coverage of basics, calculus of variations, principle of virtual work, equations of motion, more. 418pp. $5^3/_8$ x $8^1/_2$. 0-486-65067-7

Paperbound unless otherwise indicated. Available at your book dealer, online at www.doverpublications.com, or by writing to Dept. GI, Dover Publications, Inc., 31 East 2nd Street, Mineola, NY 11501. For current price information or for free catalogues (please indicate field of interest), write to Dover Publications or log on to www.doverpublications.com and see every Dover book in print. Dover publishes more than 400 books each year on science, elementary and advanced mathematics, biology, music, art, literary history, social sciences, and other areas.